Animals in Classic American Poetry

INTEGRATIVE NATURAL HISTORY SERIES
Sponsored by The Sam Houston State University
Natural History Collections
William I. Lutterschmidt, General Editor

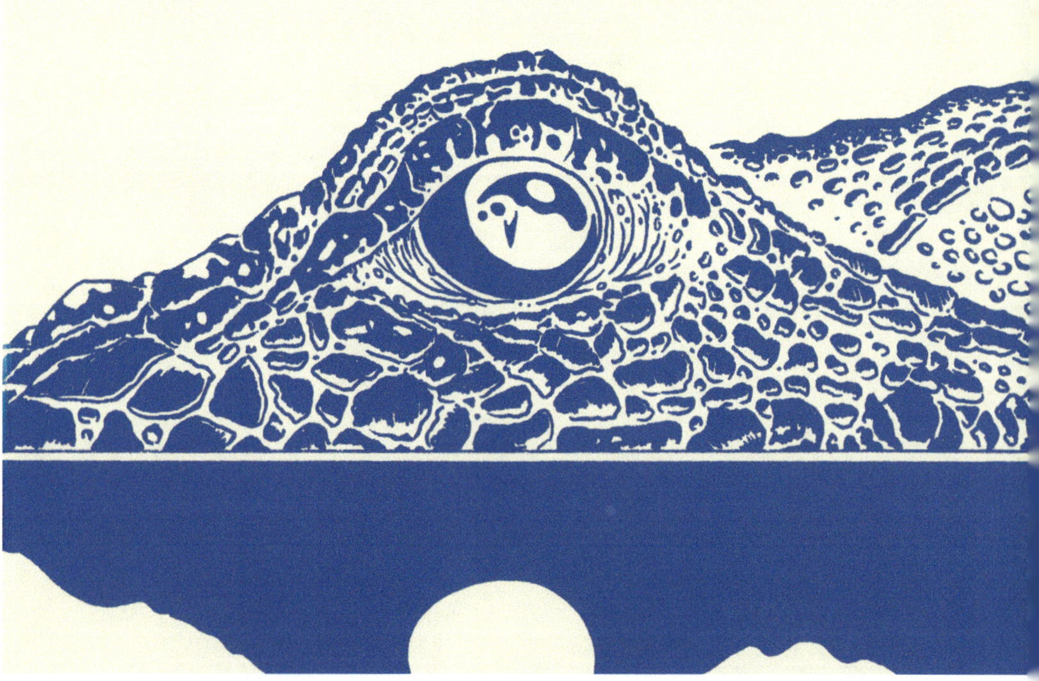

Animals in Classic American Poetry

How Natural History Inspired Great Verse

Edited by
John Cullen Gruesser

Foreword by
William I. Lutterschmidt

Texas A&M University Press • *College Station*

Copyright ©2025 by John Cullen Gruesser
All rights reserved
First edition

♾ This paper meets the requirements
of ANSI/NISO Z39.48–1992 (Permanence of Paper).
Binding materials have been chosen for durability.
Manufactured in the United States of America.

Library of Congress Cataloging-in-Publication Data

Names: Gruesser, John Cullen, 1959– editor. | Lutterschmidt, William I., writer of foreword.
Title: Animals in classic American poetry: how natural history inspired great verse / edited by John Cullen Gruesser; foreword by William I. Lutterschmidt.
Description: First edition. | College Station: Texas A&M University Press, 2025. | Series: Integrative natural history series | Includes bibliographical references and index.
Identifiers: LCCN 2024055101 (print) | LCCN 2024055102 (ebook) | ISBN 9781648433030 (paperback) | ISBN 9781648433047 (ebook)
Subjects: LCSH: American poetry—History and criticism. | Animals in literature. | LCGFT: Literary criticism. | Essays.
Classification: LCC PS310.A49 A65 2025 (print) | LCC PS310.A49 (ebook)
LC record available at https://lccn.loc.gov/2024055101
LC ebook record available at https://lccn.loc.gov/2024055102

*Several of Nature's People
I know, and they know me
I feel for them a transport
Of Cordiality*

*But never met this Fellow
Attended or alone
Without a tighter Breathing
And Zero at the Bone.*

—Emily Dickinson #1096
("A narrow Fellow in the Grass")

CONTENTS

Foreword, by William I. Lutterschmidt ix
Acknowledgments xiii

Chapter 1. "In Shady Woods I'll Sit and Sing": Human and Nonhuman Creatures in Anne Bradstreet's Poetry
Mary McAleer Balkun 1

Chapter 2. "Neither Brute nor Human": Edgar Allan Poe's Poetical Critters and Some Subsequent Revampings
Margarida Vale de Gato 23

Chapter 3. "Animality" and the "Clef of the Universes" in the Poetry of Walt Whitman
Aaron M. Moe 46

Chapter 4. Emily Dickinson's Geographic Imagination
Susan L. Roberson 67

Chapter 5. "Eden's Bad Boy": Humans and the Animal World in Melville's Poetry
Brian Yothers 92

Chapter 6. "Versed in Country Things": Animals in the Poetry of Robert Frost
Philip Edward Phillips 115

Chapter 7. Marianne Moore's Artist Animals
Heather Cass White 136

Chapter 8. Noticing Other Species—and Our Own—in Elizabeth Bishop
Calista McRae 157

Chapter 9.   Learning from Animals in Yusef Komunyakaa's Poetry
             DANIEL CROSS TURNER   179

Chapter 10. Cawing in the Dark: Avian Alterity in the Poetry
             of Joy Harjo
             THOMAS C. GANNON   201

List of Contributors   225
Index   229

FOREWORD

Animal Imagery and Natural History as Literary Inspiration

A SECOND INSPIRING CONVERSATION

I am no poet—and I know it! So, it is no surprise that I felt a level of hesitation when John Gruesser (a literary scholar) approached me (a biologist and the editor of the Integrative Natural History book series) about a new title showcasing the relationship between natural history and American poetry. Unlike our first conversation, summarized in *Animals in the American Classics: How Natural History Inspired Great Fiction* (2022), I admitted much less familiarity with poetry. However, as I reflected on the overwhelming success of our first book, I was completely confident in John's ability to organize a magnificent companion volume for the Integrative Natural History series, this time devoted to verse.

Hesitation is often triggered by the unknown. Was my unfamiliarity with poetry a missed opportunity to fully appreciate another form of literary art? As an academic scholar in science, my time is often spent reading technical and scientific reports. But this second opportunity to explore the intersections of literature and the science of natural history led to a conversation with John that inspired me to further explore what I did and did not know about poetry.

I was of course familiar with verse by Walt Whitman, T. S. Eliot, e.e. cummings, and Edgar Allan Poe. I eventually found more contemporary poets like Mary Oliver, with whose work I easily identify because it is inspired by nature—the very topic of this book. As a member of Sigma Xi (the Scientific Research Society), I remember reading Anna Lena Phillips's commentary on science and poetry in *American Scientist* where she introduces books such as *Poetry in the Wild* by Emily Grosholz and writes, "Poetry and science go way back: Over the centuries, they have occasionally gotten together, like old friends who find themselves in the same city and meet up for tea, only to head home the next day and lose touch again." So, too, with my own experiences in reading verse—I "lose touch" and feel I do not know how to fully analyze and appreciate poetry.

In our conversation, John, sensing my hesitation about the project, asked, "Do you have a favorite poem?" As a matter of fact, I do, and, not surprisingly for those who know me as a biologist, it is "A narrow Fellow in the Grass" by Emily Elizabeth Dickinson. So, John asked, "Why is it a favorite? Don't be a scientist and analyze it—just tell me why you like it." Jokingly I said, "Because it's short and simple." But seriously, the poem reminds me of being that barefoot boy bending to catch a creature that overwhelmingly fascinated me. I vividly recall seeing, as Dickinson writes, "a Whip Lash / Upbraiding in the Sun" and "stooping to secure it," only to find that "it wrinkled, and was gone." The poem brings me back to those missed catches of black racers and my declaration, "Next time I won't miss." John smiled and commented, "And there you have it—a verse by Dickinson that reminds you of your childhood and catching snakes. You have devoted your life to studying these creatures that fascinate you, and thus you appreciate Dickinson's poetic account of a narrow fellow. That's the magic of poetry: As a herpetologist, you know more about snakes than Dickinson did or ever could, and yet isn't there something that she pithily captures about them that can't be found in any scientific treatise or article, especially in her concluding phrase, 'And Zero at the Bone'?" The answer is, of course, yes. As Jean-Paul Sartre memorably puts it, "Poetry creates the myth, the prose writer draws its portrait."

As you might imagine from my informative conversation with John linking literature and the natural sciences, I began my review with Susan L. Roberson's chapter on the life and work of Emily Dickinson and the poet's ability to elegantly bring "life" to the New England landscape of the mid-1800s. As I read and reviewed each chapter in this book, I truly gained a new and profound appreciation for the insight into the natural world that only poets can provide.

I am extremely excited to add this volume to the Integrative Natural History Series and, more important, to offer a companion book to John Cullen Gruesser's *Animals in the American Classics: How Natural History Inspired Great Fiction*. As a renowned and respected literary scholar and editor, John brings together a remarkably talented group of scholarly experts on American poetry. This new volume once again elegantly showcases how animal imagery and natural history serve as literary inspiration for some of the greatest American authors.

Before you begin to explore the remarkably rich relationship between poetry and the science of natural history that only the Integrative Natural History Series can bring to you, I leave you with these words:

> A poem, in my opinion, is opposed to a work of science by having, for its immediate object, pleasure, not truth. . . . Music, when combined with a pleasurable idea, is poetry; music, without the idea, is simply music; the idea, without the music, is prose, from its very definitiveness.
>
> —*Edgar Allan Poe, "Letter to Mr. B——" (1831)*

—William I. Lutterschmidt

Works Cited

Phillips, Anna Lena. "Science and Poetry." *American Scientist* 100, no. 3 (2012): 262–68.

ACKNOWLEDGMENTS

Like *Animals in the American Classics: How Natural History Inspired Great Fiction* (Texas A&M University Press, 2022), the beautiful book you hold in your hands would not exist without the vision of Bill Lutterschmidt (and the late Brian Chapman). I must acknowledge the encouragement of other people at Sam Houston State University, namely, Jerry Cook, Will Godwin, Jacob Blevins, Chien-pin Li, Bob Donahoo, and Bernadette Pruitt, and thank the university for funding and sponsoring this book. In keeping with its predecessor, *Animals in Classic American Poetry: How Natural History Inspired Great Verse* has benefited greatly from connections resulting from my participation in professional organizations. I have known Mary Balkun, author of the essay on Anne Bradstreet (chapter 1), since the 1990s through the New Jersey College English Association (NJCEA), a vibrant organization because of her efforts and those of Kelly Shea, John Wargacki, and others at Seton Hall University, where the annual spring conference is held. Through the NJCEA I also came to know Burt Kimmelman, who heartily endorsed my decision to invite his New Jersey Institute of Technology colleague Calista McRae to contribute an essay on Elizabeth Bishop (chapter 8). I have long known Philip Phillips, who wrote about Edgar Allan Poe's "The Murders in the Rue Morgue" in the previous book and is the author of the essay on Robert Frost in this book (chapter 6), through the Poe Studies Association (PSA), of which he is currently president (for the second time). It is likewise through the PSA that I came to know Portuguese scholar and poet Margarida Vale de Gato, author of this volume's essay on Poe (chapter 2). Brian Yothers, whose contribution in the 2022 book focused on *Moby-Dick*, returns in this one with a reading of Herman Melville's poetry (chapter 5). I became acquainted with Susan Roberson, author of the essay on Emily Dickinson (chapter 4), through the American Literature Association, founded thirty-five years ago and run by Alfred Bendixen (with the able assistance of Olivia Carr Edenfield and others). I am grateful to Ed Folsom for suggesting Aaron Moe for the Walt Whitman essay (chapter 3). It was during my search for the right person to write about the Good Gray Poet that I realized Tom Gannon was the perfect choice for a piece on Joy Harjo (chapter 10), and I greatly appreciate Bernadette Russo's advice in connection with this selection. I am grateful to the people who suggested Heather White for the Marianne Moore essay (chapter 7) and to my longtime friend Malin Pereira for encouraging me to invite Daniel Cross Turner to write the one on Yusef Komunyakaa (chapter 9).

I must also thank many of the usual suspects, along with some new ones. They include Hanna Wallinger, Giulia Fabi, Tish Crawford, Carole Shaffer-Koros, Ira Dworkin, Alisha Knight, Nadia Nurhussein, Craig Werner, John Ernest, Aldon Nielsen, John Barton, JoAnn Pavletich, Charles Nelson, Tim Ungs, Matt Golden, Jim Chudy, Phil Johnson, Mike Herrmann, Joe and Edwina Murphy, Richard Katz, Bert Wailoo, Andrew Rimby, my friends (both human and canine) at Spring Lake's Mutt Beach, my uncle Chic Cullen, my Fandel cousins, my sister and brother-in-law Jenny and Mark Jansen, my children, Sarah and Jack, and my much better half, Sue.

John Gruesser
Spring Lake, New Jersey
March 2024

Animals in Classic American Poetry

CHAPTER 1

"In Shady Woods I'll Sit and Sing"
Human and Nonhuman Creatures in Anne Bradstreet's Poetry

MARY McALEER BALKUN

When Anne Bradstreet arrived aboard the *Arbella* in 1630 at what was to become the Massachusetts Bay Colony, she was entering an environment the Puritan leader William Bradford would later describe in his history of the colony, *Of Plymouth Plantation*, as "a hideous and desolate wilderness, full of wild beasts and wild men." Approximately eighteen years old and newly married, the daughter of a future governor of the colony, Thomas Dudley, and the wife of another, Simon Bradstreet, Anne Bradstreet hardly seemed suited to life on the edge of a frontier. She was educated well beyond the level of most women of her day, with a knowledge of history, philosophy, and literature, as well as Greek, Latin, French, and Hebrew. She was also well placed socially; her father was the steward to the Earl of Lincoln, and Anne lived at the estate from the age of seven to her marriage at sixteen. She was not physically strong, although she gave birth to eight children, and she was plagued by various ailments throughout her life, about which she writes in poems such as "Upon a Fit of Sickness, Anno 1632" and "For Deliverance from a Fever." Yet Bradstreet not only became the first published poet in British North America, but she was also the first to celebrate the natural beauties of the New England landscape instead of focusing exclusively on its terrors, as did so many of her contemporaries, including Bradford. Adrienne Rich, in her introduction to the Hensley edition of *The Works of Anne Bradstreet*,[1] describes her as the first American poet "to give an embodiment to American nature" (xix). However, Bradstreet's relationship to the landscape of New England and its nonhuman creatures was not without its challenges, both personal and theological.

While the environment in which Bradstreet found herself—including its geography, plant life, and indigenous inhabitants—was radically different from what she had known growing up in Northampton, England, it was New England's animal life that increasingly appeared in her poetry. Her early poems contain what can seem like pro forma lists of creatures in order to flesh out her depiction of a given region or country, but Bradstreet's later poems exhibit a more personal relationship to the animals that share her space, whether that space is local woods or the land around her home. Instead of generic inventories of animals, her later poems employ not only anthropomorphism, the attribution of human characteristics to the animals she depicts, but also zoomorphism, the rhetorical device by which humans are themselves described in terms of animal traits.[2] An examination of the poems she produced over the course of her writing life reveals Bradstreet's changing relationship to her environment as well as the means she used to reconcile her Puritan beliefs with her enjoyment of nature's gifts, including its nonhuman creatures.

Bradstreet was in many respects writing at a crucial moment in Puritan New England. Besides the trauma of settling a new land with little support and meager supplies, there were ongoing conflicts with the indigenous peoples whose land was being occupied, as well as the difficulties of the New England climate and terrain. Moreover, the faithful were being challenged by dissenters in their midst, such as Anne Hutchinson and Roger Williams, whose ideas about salvation and the authority of Puritan leadership in certain matters led both to be banished from the colony. In an environment that seemed more akin to hell than to the Eden the Puritan settlers had anticipated, new ways of thinking about the relationship between nature and the human were almost inevitable. In addition to the challenges experienced by every other Puritan colonist, given her social standing and family connections, Bradstreet was expected to serve as a model for other Puritans, especially women. The extent of these expectations is delineated in "Epistle to the Reader," a testimonial by her brother-in-law, John Woodbridge, which appeared in the front matter of Bradstreet's only book of poetry published in her lifetime, *The Tenth Muse, Lately Sprung Up*.

Having apparently taken the manuscript of her poems to England without her knowledge, Woodbridge seems compelled to explain not only that Bradstreet was the author of the verses but also that she had not written them by shirking her responsibilities either to the community or to her family. In response to those who might ask "whether it be a woman's work" (1), he observes: "If any do, take this from him that dares avow it: it is the work of woman, honoured, and esteemed where she lives, for her gracious demeanor, her eminent parts, her pious conversation, her courteous disposition, her exact diligence in her place, and discrete managing of her family occasions, and more than so, these poems are the fruit of some few hours, curtailed from her sleep and other refreshments" (1–2). While Puritans did not frown on reading or writing poetry per se, so long as it was "edifying in theme" (Rich xiii), Bradstreet herself seems to have felt the need to justify her literary activity. Even many years later, writing in her prose memoir "To My Dear Children," she explains, "I have not studied in this you read to show my skill, but to declare the Truth, not to set forth myself, but the glory of God. If I had minded the former, it had been perhaps better pleasing to you, but seeing the last is best, let it be best pleasing to you" (Hensley, *Works of Anne Bradstreet* 263). Her role as author will also be addressed in the poetry, most overtly in "The Author to Her Book," but also in a poem such as "In Reference to Her Children, 23 June 1659." Both these poems appear in the posthumously published *Several Poems Compiled with Great Wit and Learning* (1678), which included all the material in *The Tenth Muse*, as well as an additional thirteen poems (fig. 1.1).[3]

Puritan attitudes regarding the natural world have long been misunderstood and misinterpreted. While they have often been depicted as hostile to and fearful of nature, Calvinists in general and the Puritans in particular were receptive to the beauty in nature because it was God's creation. What they advised against was becoming too immersed in the natural world. According to Stephen Wolfe, because "it was so ravishing and good," nature might distract believers from "remain[ing] fixed on the eternal joys of the world to come." The place of animals in the larger culture was also changing in Bradstreet's day; whereas animals had traditionally been understood as either forms of labor or sustenance, in the early seventeenth century they gradually became more closely aligned with human beings. In her introduction to *Animalia Americana*, Colleen Glenney Boggs describes animals as "a recent invention" (21), citing Keith Thomas, whose *Man and the Natural World* "traced changing attitudes toward animals. Initially an undifferentiated part of nature that was subject to conquest and exploitation, animals only gradually gained recognition as distinct from a broader natural landscape" (Boggs 21). A contemporary of Bradstreet's, John Trapp, a Puritan chaplain, wrote that animals "are good creatures in their own nature and kind, and made to set forth the glory and magnificence of the great God." According to Philip Sampson, "This represented a significant break from the usual way of talking about animals, and it was inspired by Trapp's encyclopaedic knowledge of the Bible." Bradstreet's poetry quite literally enacts this shift: from depictions of animals as "an undifferentiated part of nature" to animals as "good creatures in their own nature and kind."[4] While some recent critics have rejected readings of Bradstreet's poetry that draw a firm line between her early and later verse, especially those that classify the early verse as public and the later verse as private,[5] others continue to recognize a certain distinction. Acknowledging these recent objections, Abram Van Engen maintains, "The thirteen additional lyrics in *Several Poems* represent something new, something not quite present in *The Tenth Muse*." This "something new" is especially evident in Bradstreet's treatment of the natural world and its nonhuman creatures.

In an early poem such as the quaternion "The Four Elements,"[6] for example, animals appear primarily in the form of the lists referred to previously. In the "Earth" section of the poem, traveling from cities to mountains and finally to hills and plains, the speaker enumerates the creatures that populate the global landscape in order to prove the superiority of Earth to the other elements: There are "Lions of Numidia," "Panthers and . . . Leopards of Libia," the "Behemoth and rare found Unicorn," and the Hyaena (lines 177–81). And these are just a few of the "Thousands in woods and plains / both wild and tame" that the speaker "list[s] now none to name" (lines 183–84). The one point in the poem at which there might have been a moment of connection between the human and the nonhuman comes when "the fawning dog" tries to get the speaker to "commend" his "trust and valour," but the possibility is rejected when the speaker explains rather brusquely that "time's too short and precious so to spend" on such things (lines 185–88).

Bradstreet uses the same cataloging technique in the "Water" section of the poem, this time to illustrate the fact that all creatures—not just human ones—are affected similarly by drought:

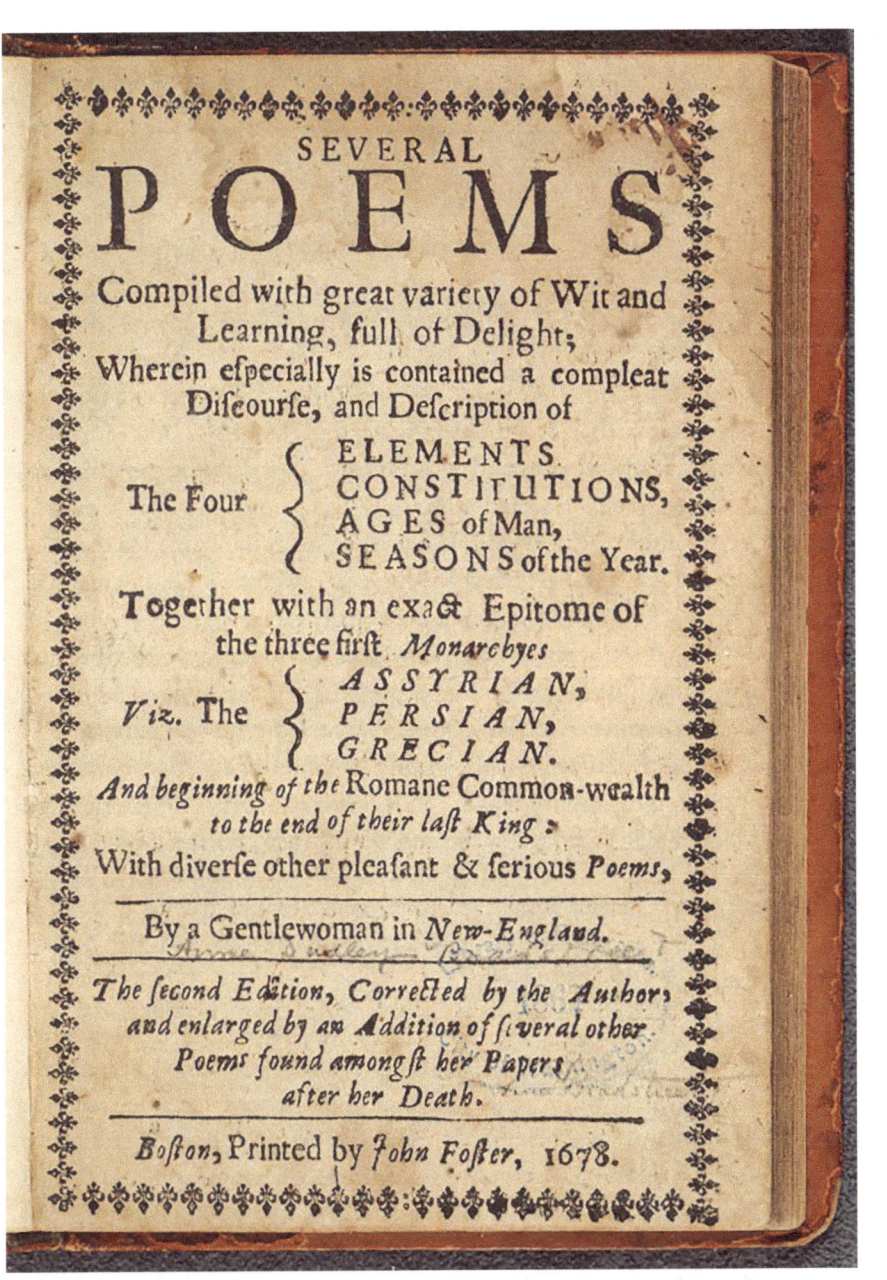

Figure 1.1. Title page of *Several Poems Compiled with Great Variety of Wit and Learning* (Boston: John Foster, 1678). Rare Book and Special Collections Division, Library of Congress.

> *Thy bear, thy tiger and thy lion stout,*
> *When I [Water] am gone, their fierceness none needs doubt*
> *Thy camel hath no strength, thy bull no force*
> *Nor mettal's found, in the courageous horse*
> *Hinds leave their calves, the elephant, the fens*
> *The wolves and savage beasts, forsake their dens*
> *The lofty eagle, and the stork fly low,*
> *The peacock and the ostrich, share in woe (lines 282–89)*

But it is only the human creatures who are capable of understanding the causal relationship between the lack of water and the lack of other forms of sustenance—"bread and wine, and pleasant fruits"—and so the thirsting person will look for water "in river and in well / His [drought's] deadly malady I might expel" (lines 292, 294–95). This privileging of the human is repeated when the poet describes the effects of too much water on God's creatures: while human and nonhuman animals are equally impacted, only the human ones can complain "Of rotten sheep, lean kine, and mildewed grain" because "with my wasting floods and roaring torrent / Their cattle, hay, and corn I [water] sweep down current (lines 365–67). The pattern is evident again in the "Air" section, where contagions of various kinds mean "That birds have not 'scapt death as they have flown" and "Of murrain, cattle numberless did fall," but it is humans who are aware enough to "[fear] destruction epidemical" (lines 464–66). Thus, while animals may be "an undifferentiated part of nature," to use Thomas's language, and as much a part of the natural world as their human counterparts, it is the suffering of God's human creatures that is given precedence time and again in these early poems.

One reason for Bradstreet's lack of emotional engagement with animals in "The Four Elements" may be that the creatures she names are largely a reflection of her education rather than her direct experience. The creatures associated with "Air," for instance, include the ostrich, phoenix, stork, and crane, as well as the more common partridge, thrush, wren, and lark. Bringing together the known and the unknown in this way, without distinctions for those with which she would have been familiar, creates the effect of a natural world discrete from the human creature observing it, where animals occupy the same space and are part of God's creation but their relationship to the human (and vice versa) is essentially nonexistent. The animals have more in common with one another than with her. This emotional distancing is an example of what John Gatta identifies as "the Puritan principle of 'weaned affections' [which] stipulated that human beings . . . must cultivate enough detachment from the world to understand that they had no permanent habitation on earth" (46). Thus, while the poem establishes the overall unity of nature, which is in keeping with Puritan theology, it also reinforces the idea that one should not allow oneself to become invested in the things of this world, including nature.

Yet even in these early, more formal poems, there are glimmerings of a more personal relationship to the creatures that share Bradstreet's space. These instances generally take the form of anthropomorphism. In these moments of connection, the creatures exhibit domestic and maternal traits, but they are also creatures she would have known firsthand (fig. 1.2). Several such moments can be found in the "Spring" section of "The Four Seasons," another of Bradstreet's quaternions. For example, about March and the emergence of spring, she writes:

> *The nightingale, the blackbird and the thrush*
> *Now tune their lays, on sprays of every bush.*
> *The wanton frisking kid, and soft-fleeced lambs*
> *Do jump and play before their feeding dams,*
> *The tender tops of budding grass they crop (lines 34–38)*

From an imagined tuning of voices to the cheerful image of young goats and sheep amusing their indulgent mothers, the speaker creates a world where—unlike that in "The Four Elements"—the creatures, like their human counterparts, "joy in what they have, but more in hope" (line 39).

This delight and engagement continue into April:

> *The fearful bird his little house now builds*
> *In trees and walls, in cities and in fields.*
> *The outside strong, the inside warm and neat;*
> *A natural artificer complete.*
> *The clocking hen her chirping chickens leads*
> *With wings and beak defends them from the gledes.*[7] *(lines 56–61)*

Then in May the speaker observes "the busy, witty, honey-bee / Whose praise deserves a page from more than me" (lines 70–71). These are crea-

8 | CHAPTER 1

Figure 1.2. *Wood Thrush, 1. Male 2. Female (Common Dogwood).* Drawn by J. J. Audubon. J. J. Audubon, *The Birds of America: From Drawings Made in the United States and Their Territories,* 1841. New York Public Library Digital Collections.

tures that create safe and cozy spaces for their offspring and are witty as well as busy. Bradstreet's descriptions emphasize the maternal aspects of these fellow creatures, thereby making them a more authentic reflection of herself. Finally, like the poet, they are "natural artificer[s]," occupying a median position between nature and the ultimate Creator, not challenging God's creation but emulating and thereby glorifying it.

In fact, the depiction of animals in "The Four Seasons," while still similar to the lists of creatures seen in "The Four Elements," anticipates Bradstreet's treatment of animals in the later, lyrical poems, where her connection to the creatures of the earth is not only more nuanced but also more subjective. This deeper engagement with the natural world, most often observed in regard to Bradstreet's poem "Contemplations," can be traced to a practice of reflection known as "meditation on the creatures."[8] According to Robert Daly, the Puritans saw God's will laid out in two books, the book of scripture and the book of nature, and "reading this book of the world was as common a religious practice among Puritans as it was among other Christians in the Middle Ages" (370). Gatta makes the same connection, arguing that "Contemplations" can be read "as a certain type of nature poem: a meditation on the creatures founded in Christian tradition but stirred by encounter with the New World" (42). However, meditation on the creatures could also be a humbling experience, as the Puritan theologian William Perkins explains in *Treatise of Man's Imagination (1607):* "by comparing ourselves with the brute creatures, we may learn to humble ourselves and to be abashed when we see them which were made to serve us, to go before us in obedience to the laws of our creator" (qtd. in Watson 1127). Meditation on the creatures was therefore expected to provide the same benefits as meditation on the scriptures: examples of behavior to be emulated or shunned, deeper insights into the self, as well as a deeper connection to the Creator of all things.[9] Since meditation on the creatures is a form of spiritual practice, it should not be surprising that the poems where Bradstreet more deeply engages with animals are also those that more fully reveal her personal struggles with her faith. These poems reveal a tension between what Bradstreet believes she should be feeling, thinking, or experiencing in a given moment and what is actually transpiring. Such pressures are in full display in "Contemplations," one of her most admired poems.

On an autumn day, Bradstreet decides to take a walk in the local woods, and the moment becomes an opportunity for her to reflect on the beauties of the creation and her own place within it. The rising sun warms the landscape and its inhabitants—"Birds, insects, Animals with Vegative" (line 34)—and motivates the speaker

> *To sing some song my mazed Muse thought meet*
> *My great Creator I would magnify,*
> *That nature had thus decked liberally (lines 54–56)*

However, hearing other creatures whose voices rival her own, she immediately declares her "imbecility!" (line 57) and chooses to listen instead:

> *I heard the merry grasshopper then sing,*
> *The black-clad cricket bear a second part;*
> *They kept one tune and played on the same string,*
> *Seeming to glory in their little art. (lines 58–61)*

In comparison to these supposedly "abject" creatures (line 62), she "as mute, can warble forth no higher lays" (line 64). These singing creatures reappear in stanzas 24 to 30, offering a striking and emotional counterpoint to Bradstreet's frail human condition. The section begins with the speaker observing a "stealing stream . . . / Which to the longed-for ocean held its course" (lines 149–50). Unlike her own path in life, the stream encounters nothing that might impede it, leading the speaker to remark:

> *"O happy flood," quoth I, "that holds thy race*
> *Till thou arrive at thy beloved place,*
> *Nor is it rocks or shoals that can obstruct thy pace." (lines 153–55)*

Besides the lack of obstacles, the stream is, unlike the speaker, not solitary in its journey:

> *Nor is't enough that thou alone may'st slide,*
> *But hundred brooks in thy clear waves do meet,*
> *So hand in hand along with thee they glide*
> *To Thetis house, where all embrace and greet. (lines 156–59)*

Her reflection on the stream becomes an opportunity to direct her attention to heaven, wishing that she too could "lead my rivulets to rest / So may we press to that vast mansion, ever blest" (lines 161–62).

The speaker then considers the fish that populate the waters (fig. 1.3), which move from salt to fresh water depending on the season and where they "think best to glide" (line 165), traveling "To unknown coasts to give a visitation" (line 166). Even more important, they reproduce without the concerns of their human counterparts, unaware of their good fortune in not having to question or wonder:

> *In lakes and ponds, you leave your numerous fry;*
> *So nature taught, and yet you know not why,*
> *You wat'ry folk that know not your felicity.*[10] *(lines 167–69)*

Figure 1.3. Sample engraving of an Atlantic salmon (*Salmo salar*, Linnaeus, 1758). From Marcus Elieser Block, *Oeconomische Naturgeschichte der Fische Deutschlands*, published 1782–85. Wikimedia Commons.

This apparent envy of blissfully ignorant animal life continues in stanza 25, as the speaker describes "how the wantons frisk to taste the air / Then to the colder bottom straight they dive" (lines 170–71), while other fish "forage o'er the spacious sea-green field / And take the trembling prey before it yield" (lines 174–75). Suddenly, absorbed in these reflections, she is joined by "The sweet-tongued Philomel" (or nightingale) who

> *. . . chanted forth a most melodious strain*
> *Which rapt me so with wonder and delight,*
> *I judged my hearing better than my sight,*
> *And wished me wings with her a while to take my flight. (lines 180–83)*

Not content to watch and listen, the poet yearns to become a bird herself. Imagining the existence of her winged companion, she directly addresses a creature that she thinks

> *fears no snares,*
> *That neither toils nor hoards up in thy barn,*
> *Feels no sad thoughts, nor cruciating cares*
> *To gain more good, or shun what might thee harm*
> *Thy clothes ne're wear, thy meat is everywhere,*

> *Thy bed a bough, thy drink the water clear,*
> *Reminds not what is past, nor what's to come dost fear. (lines 184–90)*

As with the fish, this bird also does not lead a solitary existence. It has a "feathered crew" (line 192) that joins it in song and then follows it "into a better region / Where winter's never felt by that sweet airy legion" (lines 196–97). In contrast to the fish and the birds, and even to the earlier grasshopper and cricket, however, humans suffer from an array of problems and insecurities. In fact, the speaker's primary emotion when regarding many of the creatures she encounters is envy, a desire to be, like them, free from human cares and consciousness. Her reflections simply confirm that

> *Man at the best a creature frail and vain,*
> *In knowledge ignorant, in strength but weak,*
> *Subject to sorrows, losses, sickness, pain,*
> *Each storm his state, his mind, his body break,*
> *From some of these he never finds cessation,*
> *But day or night, within, without, vexation,*
> *Troubles from foes, from friends, from dearest, nearest relation. (lines 198–204)*

Despite being a "lump of wretchedness, of sin and sorrow" and a "weather-beaten vessel wrackt with pain"—a line that seems to speak directly to Bradstreet's own poor health—humans also fail to place their hope in "an eternal morrow" (line 208). And even to someone who does not experience much actual suffering in life, "sad affliction comes and makes him see / Here's neither honor, wealth, nor safety" because "Only above is found all with security" (lines 223–25). The poem thus ends where it should, according to Puritan mores, with a reaffirmation of the life and world beyond this one. However, there is something distinctly unsatisfying about this last section of the poem, where Bradstreet turns away from the world in which she has been immersed—with its beauties, its joys, and its absence of cares—and redirects her attention to the plight of God's human creatures.

Many critics have commented on the way Bradstreet's poetry reflects her effort to remain focused on salvation and the afterlife as opposed to succumbing to the pull of earthly things—loved ones, cherished objects, safety, and security.[11] Poems such as "Before the Birth of One of Her Children" and "Verses Upon the Burning of Our House July 10, 1666" exhibit this overtly, with Bradstreet worrying in the first about the fate of her children if she should die in childbirth, but also recognizing that "All things within this fading world hath end" (line 1), and in the second mourning the loss of her earthly possessions but in the end accepting that her "hope and treasure lies above" (line 58). This pattern is evident in her more formal verses as well, such as "The Vanity of All Worldly Things." Originally (and intriguingly) titled "The Vanity of All Worldly Creatures," the poem not only argues for the transience of the material world, but it is another instance of Bradstreet using animals as a counterpoint to the human condition. Here, the path that leads to salvation is one "no vulture's eye hath seen / Where lion fierce, nor lion's whelps have been" (lines 36–37); it is one intended only for God's human creatures.

The challenge to find equanimity and refocus on the afterlife plays out much differently in "Contemplations," however; while the poem traces Bradstreet's spiritual struggle in a deeply personal way, it does so *through* her engagement with the natural world, especially its animals. If "Contemplations" is indeed the place in Bradstreet's body of work where her ecological sensibility is most evident, as some critics maintain,[12] then this could also be why her ultimate rejection of what has been spoken of in the first twenty-eight stanzas of the poem seems obligatory rather than deeply felt. Her recitation of the frailties of the human, the lure of vanity, and the ravages of time does not contain new or original observations; rather, they are traditional Puritan reflections on the material world.[13] Although Bradstreet may not be willing at this stage to fully align herself with the natural world and move beyond the strictures of traditional meditation on the creatures, thereby invoking a vision of a unified creation, she has no such qualms in two later poems. In "Another (II)" and "In Reference to Her Children, 23 June 1659,"[14] we see a poet who ultimately decides to fully embrace the animal other in creative and even amusing ways. As Danielle Sands observes, "We use animal others as means to map our internal landscapes, projecting the strangest, most disturbing parts of our nature elsewhere in the hope of stabilizing the turmoil within. We use animal figures not because we know ourselves and understand the ways that we differ from animals, but rather because, for all our protestations otherwise, our identity slips away from the structures we create to account for it" (459). Bradstreet's engagement with nonhuman creatures is overt in these two poems, both of which are deeply personal: one is a love poem to her husband away on business, and the other is about her children growing up and leaving home. It is also in these poems that Bradstreet's use of anthropomorphism is abandoned in favor of zoomorphism; she shifts from simply reflecting on animals as symbols for the human condition to an active identification with them, merging the human and the nonhuman in order to arrive at her own means of reconciling this world and the next.

"Another (II)" is the third of three poems Bradstreet wrote for her absent husband expressing how much she misses him. The first, "A Letter to Her Husband, Absent Upon Public Employment," is reminiscent of John Donne's "A Valediction Forbidding Mourning," with the speaker depicting herself and her absent husband as a single unit: "Flesh of thy flesh, bone of thy bone / I here, thou there, yet both but one" (lines 25–26). In "Another (I)" she calls on Phoebus (Apollo, the sun god in Greek mythology) to relay her "griefs" (line 4) to her missing partner, figuring herself as a "widowed wife" (line 10) in mourning due to his absence. She directs the sun to "Tell him here's worse than a confused matter / His little world's a fathom under water" (lines 33–34). In the final poem of the set, however, "Another (II)," Bradstreet casts herself and her husband in the roles of several animal pairs: deer, turtledoves, and mullets (fig. 1.4). This is also the most fanciful poem of the three, with Bradstreet employing homophones and double entendres in playful ways. The speaker is first the "loving hind that (hartless) wants her deer" (line 1), who "Scuds through the woods and fern with harkening ear" (line 2) searching for her mate, to no avail. Yet the repetition of "deer"/"dear" in these lines gives a lighter cast to her longing:

> *Her dearest deer, might answer ear or eye;*
> *So doth my anxious soul, which now doth miss*
> *A dearer dear (far dearer heart) than this. (lines 4–6)*

She and her absent husband are then twin turtledoves, with her half of the pair akin to the actual dove that makes "most uncouthly bemoan" in its solitary state (line 10):

> *Even thus do I, with many a deep sad groan,*
> *Bewail my turtle true, who now is gone,*
> *His presence and his safe return still woos,*
> *With thousand doleful sighs and mournful coos. (lines 13–16)*

CREATURES IN ANNE BRADSTREET'S POETRY | 15

Figure 1.4. *Common Deer. The Cabinet of Natural History and American Rural Sports, with Illustrations* (Philadelphia: J. & T. Doughty, 1830), pl. I. Wikimedia Commons.

Finally, she is a mullet, "that true fish" (line 17), who, seeing her mate captured, flings herself upon the shore "there for to die" (line 19) rather than live without him. As she does in the poem that precedes it, "Another (I)," Bradstreet compares her lonely state to widowhood, writing, "I have a loving peer, yet seem no wife" (line 22). And while the first sections of this poem engage in quite traditional forms of analogy, in the final section Bradstreet and her husband *are*, at least for a moment, hart and hind, turtledoves, and mullets—"Return my dear, my joy, my only love / Unto thy hind, thy mullet, and thy dove" (lines 25–26)—hopefully to remain, like their animal counterparts, together until death. The line between human and animal wavers, blurs, and then collapses, as Bradstreet finds in these creatures a constancy and emotional connection to a mate like her own.

In "Another (II)" Bradstreet moves easily from anthropomorphism to zoomorphism, using extended similes to paint a vivid picture of each animal pair. She also uses local creatures with which she would have been familiar and very likely would have observed as they moved through the landscape: running through woods, sitting on boughs, gliding through water. As in some of the earlier poems discussed previously, the imagery is domestic in nature: when her husband returns, they will eat together, live together, and move through the world side by side:

Together at one tree, oh let us browse,
And like two turtles roost within one house,
And like the mullets in one river glide (lines 29–31)

But the speaker is also trapped within her domestic role, just as her husband is in his civic one, neither able to reach the other: "But worst of all, to him can't steer my course / I here, he there, alas, both kept by force" (lines 23–24). "Another (II)" is a love poem that draws on images of animal constancy, and even innocence, rather than on other images of human or divine love; as a good Puritan, however, Bradstreet could certainly maintain that since God made these loving creatures, they represent his love as well (fig. 1.5). The single concession to the world beyond this one comes when she writes, "Let's still remain but one, till death divide" (line 32), acknowledging that worldly love and unity have their limits.

An even more dramatic and imaginative coalescing of the human and the animal occurs in "In Reference to Her Children, 23 June 1659," which is also Bradstreet's most marked use of zoomorphism. Figuring her eight children as "eight birds hatcht in one nest" (line 1) with herself as the mother bird who "nursed them up with pain and care" (line 3), the speaker describes each of her avian offspring in turn as they grow up and leave, an experience both human and nonhuman creatures share. There is the "Chief of the brood" (line 7) who traveled "To regions far and left me quite" (line 8); the second and third of her offspring marry and set up new homes of their own; the fourth "to the Academy flew / To chat among that learned crew" (lines 27–28); the fifth is beginning to find his way, "And as his wings increase in strength / On higher boughs he'll perch at length" (lines 35–36). Finally, while three "birds" still remain in the nest, the speaker recognizes that they too will eventually leave: "Or here or there, they'll take their flight / As is

Figure 1.5. *Turtle-Dove.* Lithograph by John Gould, issued 1862–73. New York Public Library Digital Collections.

ordained, so shall they light" (lines 39–40). Whereas in prior poems the speaker is aligned with animals only to pull back from that integration, in this case Bradstreet sees the analogy—for herself and for her children—through to the end:

> *If birds could weep, then would my tears*
> *Let others know what are my fears*
> *Lest this my brood some harm should catch*
> *And be surprised for want of watch*
> *Whilst pecking corn and void of care*
> *They fall un'wares in Fowler's snare;*
> *Or whilst on trees they sit and sing*
> *Some untoward boy at them do fling,*

18 | CHAPTER 1

> *Or whilst allured with bell and glass*
> *The net be spread and caught, alas.*
> *Or lest by Lime-twigs they be foiled,*
> *Or by some greedy hawks be spoiled. (lines 41–52)*

Her protecting arms are "wings" that "kept off all harm" (line 58) while they were young; in her later years, she claims, "In shady woods I'll sit and sing / And things that past, to mind I'll bring" (lines 69–70). The whimsical descriptions of her children as birds—their coloring, their traits—as well as the evocative descriptions of their behavior—moving to higher boughs, flying southward, losing their "down"—imply an author who has paid a great deal of attention to the actual creatures. As in "Contemplations" and "Another (II)" the speaker connects to other creatures through domestic images: bearing children, making a safe home for them, and ensuring they are fed and cared for. Although the poem seems to be about the birds the speaker has raised, it is in reality about the poet herself, the plight of motherhood, and her own mortality. Where the poet was "Once young and pleasant" (line 72) as her children are now, she has spent her life bringing them safely to adulthood only to have them move off to begin their own lives, where the pattern begins again. However, this mother is also a poet, one who sings to her "birds" while they are with her and who will continue to sing during her remaining years:

> *Meanwhile my days in tunes I'll spend,*
> *Till my weak lays with me shall end.*
> *In shady woods I'll sit and sing*
> *And things that past, to mind I'll bring. (lines 67–70)*

Poetry thus becomes a way to preserve the past and herself, since even after her departure her offspring will have her poems to remember her by. Bradstreet also suggests, for the first time, that her own song making will find a place in heaven:

> *My age I will not once lament*
> *But sing, my time so near is spent,*
> *And from the top bough take my flight*
> *Into a country beyond sight*

Where old ones instantly grow young
And there with seraphims set song. (lines 73–78)

She will finally join her song with those of the angels. In addition, her children will "In chirping languages" (line 83) tell their own offspring about her, thereby keeping her alive. Bradstreet can therefore take comfort in three types of immortality: not only eternal life in heaven and her children's memories of her passed on to their descendants but also her poetry itself.

Bradstreet's identification with the creatures in "In Reference to Her Children" enables her to represent both the joys and sorrows of motherhood with some psychological distance as well as with a touch of whimsy. As she has pointed out in earlier poems, nonhuman creatures do not have the awareness that humans have; they are blissfully ignorant and do not know enough to question their purpose or their eventual fate. In this poem, however, Bradstreet envisions a mother bird who realizes what is happening and what will happen and feels pain and sadness. In both "In Reference to Her Children" and "Another (II)," identification with the creatures enables her to provide a richer view of a unified nature, where human and nonhuman creatures share certain experiences, and perhaps even a similar fate.

Along with the typical Puritan conflict between this world and the next, Bradstreet's poetry captures her struggles with her responsibilities as a Puritan wife and mother in a challenging new environment. However, in her later poetry particularly, nature comes to represent a space of integration. It is not only free from domestic cares and restrictions, but it is also the place where, through her engagement with nonhuman creatures, she can work through these tensions and imagine a resolution unique to her. By the time she writes the poem to her children in 1659, she apparently feels confident enough to create a space where the resonances between the human animal and the nonhuman animal can finally take center stage, affirming her affinity with the totality of the creation.

Notes

1. Throughout this chapter I follow Hensley's punctuation and capitalization of the poems.

2. For more on zoomorphism, see Kári Driscoll and Eva Hoffman, "Introduction: What Is Zoopoetics?"

3. For a detailed discussion of the issues around the publication of Bradstreet's poetry, both *The Tenth Muse* and the posthumously published *Several Poems Compiled with Great Wit and Learning*, see Margaret Olofson Thickstun, "Contextualizing Anne Bradstreet's Literary Remains: Why We Need a New Edition of the Poems."

4. In "Advertising the Domestic: Anne Bradstreet's Sentimental Poetics," Abram Van Engen notes that "this newer approach, while useful in directing attention to continuities in the poetry, should still be tempered by the general shift in style and subject matter that does seem to occur."

5. See, for example, Elizabeth Ferszt and Ivy Schweitzer's introduction to the special issue of *Women's Studies* about Bradstreet (286).

6. A quaternion is a poem in which a single theme is explored in four parts.

7. A glede is a type of bird. According to the *Oxford English Dictionary*, in addition to the "kite," "the name is also locally applied to other birds of prey, as the buzzard, osprey, and peregrine falcon" (OED "glede").

8. This practice is alternatively referred to both as "meditation from the creatures" and "meditation on the creatures." I have chosen to use the latter in this chapter.

9. In addition to Daly and Gatta, see Barbara Keifer Lewalski's *Protestant Poetics and the Seventeenth-Century Religious Lyric* for more on this practice.

10. Gatta, in *Making Nature Sacred*, makes a strong case for the fish being described here as an Atlantic salmon (48).

11. See, for example, William J. Scheick's *Authority and Female Authorship in Colonial America* and Carrie Galloway Blackstock's "Anne Bradstreet and Performativity."

12. See, for example, Daly; and Gatta.

13. In "Bliss Lost, Wisdom Gained: Contemplating Emblems and Enigmas in Anne Bradstreet's 'Contemplations'" Michael Ditmore makes a similar point about the poem, observing that "its theological meaning does not fit neatly into normative categories and is difficult to tease out; to complicate matters, it also raises the possibility that the poem's actual theology is at variance with its imputed Calvinist Puritan contexts" (44).

14. "Another (II)" is the title typically used for this poem. Thickstun notes that we cannot even be sure the titles of some poems are Bradstreet's and that poems such as this one were likely included in letters to her husband so would not have been titled at all (409).

Works Cited

Blackstock, Carrie Galloway. "Anne Bradstreet and Performativity." *Early American Literature*, December 1997, 222–49.

Boggs, Colleen Glenney. *Animalia Americana*. New York: Columbia University Press, 2013.

Bradford, William. *Of Plymouth Plantation*. Boston: Wright and Potter, 1898. Project Gutenberg. Accessed October 10, 2022. https://www.gutenberg.org/files/24950/24950-h/24950-h.htm.

Daly, Robert. "Reading God's Book of the World." In *A Companion to American Poetry*, edited by Mary McAleer Balkun, Jeffrey Gray, and Paul Jaussen, 369–78. Hoboken, NJ: Wiley-Blackwell, 2022.

Ditmore, Michael. "Bliss Lost, Wisdom Gained: Contemplating Emblems and Enigmas in Anne Bradstreet's 'Contemplations.'" *Early American Literature* 41, no. 1 (2007): 31–72.

Driscoll, Kári, and Eva Hoffman. "Introduction: What Is Zoopoetics?" In *What Is Zoopoetics? Texts, Bodies, Entanglement*, edited by Kári Driscoll, Eva Hoffman, and Marcel Beyer, 1–13. London: Palgrave Macmillan, 2018.

Ferszt, Elizabeth, and Ivy Schweitzer. "Introduction." *Women's Studies* 43, no. 3 (2014): 285–89.

Gatta, John. *Making Nature Sacred: Literature, Religion, and Environment in America from the Puritans to the Present*. Oxford: Oxford University Press, 2004.

Hensley, Jeannine. "Anne Bradstreet's Wreath of Thyme." In *The Works of Anne Bradstreet*, edited by Jeannine Hensley, xxiii–xxxix. 1967. Reprint, Cambridge, MA: Belknap Press of Harvard University Press, 2010.

———, ed. *The Works of Anne Bradstreet*. 1967. Reprint, Cambridge, MA: Belknap Press of Harvard University Press, 2010.

Lewalski, Barbara Keifer. *Protestant Poetics and the Seventeenth-Century Religious Lyric*. E-book. Princeton, NJ: Princeton University Press, 1979.

Rich, Adrienne. "Anne Bradstreet and Her Poetry." In *The Works of Anne Bradstreet*, edited by Jeannine Hensley, ix–xxii. 1967. Reprint, Cambridge, MA: Belknap Press of Harvard University Press, 2010.

Sampson, Philip. "What on Earth Are Animals For?" SARX: For All God's Creatures. Accessed October 9, 2022. https://sarx.org.uk/articles/christianity-and-animals/what-on-earth-are-animals-for/.

Sands, Danielle. "Religion." In *The Edinburgh Companion to Animal Studies*, edited by

Lynn Turner and Undine Sellbach, 459–74. Edinburgh: Edinburgh University Press, 2018.

Scheick, William J. *Authority and Female Authorship in Colonial America*. Lexington: University Press of Kentucky, 2015.

Thickstun, Margaret Olofson. "Contextualizing Anne Bradstreet's Literary Remains: Why We Need a New Edition of the Poems." *Early American Literature* 52, no. 2 (Spring 2017): 389–422.

Van Engen, Abram. "Advertising the Domestic: Anne Bradstreet's Sentimental Poetics." *Legacy* 28, no. 1 (2011): 47–68. https://www.jstor.org/stable/10.5250/legacy.28.1.0047.

Watson, Robert N. "Protestant Animals: Puritan Sects and English Animal-Protection Sentiment, 1550–1650." *ELH* 81, no. 4 (December 2014): 1111–48. https://www.researchgate.net/publication/290519691_Protestant_Animals_Puritan_Sects_and_English_Animal-Protection_Sentiment_1550–1650.

Wolfe, Stephen. "The Worldly Poetry of the Puritans." Mere Orthodoxy, July 1, 2016. https://mereorthodoxy.com/worldly-poetry-puritans/.

Woodbridge, John. "Epistle to the Reader." In *The Works of Anne Bradstreet*, edited by Jeannine Hensley, 1–2. 1967. Reprint, Cambridge, MA: Belknap Press of Harvard University Press, 2010.

CHAPTER 2

"Neither Brute nor Human"
Edgar Allan Poe's Poetical Critters and Some Subsequent Revampings

MARGARIDA VALE DE GATO

In my class discussions of Edgar Allan Poe's "The Raven," while glossing the part where the bird perches upon the bust of Pallas, I refer as a rule to the "effect of contrast" claimed by the metatext "The Philosophy of Composition" (Poe, *Poe's Critical Theory* 68). Drawing on the multimodal perceptions of visual grammar furnished by Gunther Kress and Theo van Leewen, I point to this contrast as an inversion of the top-bottom compositional structure (193–98), with the animal motif placed on the higher, ideal plane and the intellectual on the lower, baser plane. An aesthetic figuration of reason (the marble Pallas) is thus overshadowed by the primitivism of the wild, dark animal. In the past semester, at this point, a student raised his hand and said that he had understood it differently; having watched online a *Nature* documentary about the intelligence of ravens,[1] he thought the bust of Pallas was meant to enhance the bird's status as sage, or, as it is called in the poem, "prophet" (Poe, *Poems* 368). In fact, "The Philosophy of Composition" also offers us the contention that "the bust was absolutely *suggested* by the bird" (*Poe's Critical Theory* 68; emphasis in the original). Perhaps by a turn of phrase that is also a slip of authority, Poe here delegates the agency of composition to the bird and thereby challenges intentionality as a human trait separating us from other animals.[2] Furthermore, the human dominion of intelligible speech is problematized in "The Raven" as elsewhere in Poe's work, for example, in "The Murders in the Rue Morgue" in which the sounds of an orangutan are mistaken for those of a human murderer, and the passage in *Arthur Gordon Pym* where the shriek of a gull is mistaken for the admonition of a mariner. In the case of "The Raven," we may turn again to "The Philosophy of Composition" for "the idea of a *non*-reasoning creature capable of speech" (*Poe's Critical Theory* 65). This idea, as Colleen Boggs asserts, unsettles the properties of the human since "Poe unhinges reason from speech, and abstracts tone, a physical quality of poetry, and affect (melancholy) as the somatic origins, and aims, of literary work" (127). Poetry and animal sounds, then, are set loose from the exclusive domain of human reason and become sensorial phenomena upsetting the mental realm, which is also denied the sole control of speech.

But if Poe really did want to detach speech from reason, why does the bird sit on the head of Athena? Could this be a recollection of a line we find in the jocular juvenile poem "Oh, Tempora! Oh, Mores!": "Philosophers have often held dispute / As to the seat of thought in man and brute" (*Poems* 11), later expounded on in the essay "Instinct vs. Reason: A Black Cat"?[3] Moreover, how far is the bird used to explore "under-current[s] . . . of meaning" (*Poe's Critical Theory* 70), and how much do these relate to forms of symbolic communication that actually occur across species? Such forms have been labeled "transspecific" by Michael Ziser, who also uses the term "zoosemiotics," or the study of the use of signs among animals, to support his interpretation of "The Raven" as a play on the ability of (de)coding between human and animal (11).

What if we look at Poe's masterpiece as being mostly about a raven, as its title indicates? Indeed, my student's question, along with the proposal to write this chapter and read Poe through the lens of animal studies, opened perspectives that I pursue here. I want to highlight how animals in Poe's poems do not matter so much because they are symbolic or impenetrable but rather because they cross thresholds, combining unexpected matters (e.g., marble and plumage, in the case of "The Raven") and beings (e.g., angels and vermin in "The Conqueror Worm"), often in liminal and dynamic forms. I first examine how animals figure in Poe's poetry both as stand-*ins*, as if they were to represent something, and as stand-*fors*, as if they could be agents of social interaction alongside humans—that is, representative animals, as it were, fulfilling *something* of an agent-minded role. The arbitrariness and changeability of that *something* might indicate a slippage of the prerogatives of human and animal, especially in terms of language ability. Second, I investigate Poe's hybrid and sometimes bionic "critters," a term used by Donna Haraway to encompass "a motley crowd of lively beings including microbes, fungi, humans, plants, animals, cyborgs, and aliens" (330), bound to be entangled with each other in a time that exceeds our own.[4] Finally, I look at how later artists have recast these animate beings, contributing to the enrichment of the *com-post*—another of Haraway's critical neologisms, which stresses entanglement and intersubjectivity (*com*) in our *post*-age (*Staying* 11, 31)—of composition.

Represented and Representing Animals in Poe's Poetry: Ambivalent Symbolization and Obscure-Source Mythology

What is interesting in Poe's poetic use of animals, most of them birds, is that they are not easily classified as representations vis-à-vis the human,[5] but rather they seem chosen by some quality that entails a contradiction in the terms of comparison. In the 1831 version of "Tamerlane," a "vampire-bat" is the metaphor chosen for the edginess of Despair, evoking both restlessness and stasis: "Despair, the fabled vampire-bat, / Hath long upon my bosom sat, / And I would rave, but that he flings / A calm from his unearthly wings" (*Poems* 46). This "fabled" creature—not really a bird, but a composite mammal—will be omitted from the version of "Tamerlane" published in *The Raven and Other Poems* (1845), perhaps to avoid a weak echo of the powerful image of the avian beak piercing the speaker's heart in "The Raven." But the invocation of a rapacious bird feasting upon suffering but worthy flesh also appears in "Science," a rare sonnet, with the lines: "Why preyest thou upon the poet's heart / vulture, whose dreams are dull realities?" (*Poems* 90).

Even if underrepresented in terms of species and rarely detailed, animals in Poe's poetry are not meant to belittle or praise human qualities, nor are they straightforwardly allegorical, for they do not offer a translucent glimpse of the idea they stand for but complicate it. In "Tamerlane" and "To One in Paradise," eagles seem to stand for love, but they appear as preternatural omens, with "snowy wings" and "light'ning eye" unsparing to "mote" or "fly" (*Poems* 61), or "stricken" and thus no longer able to "soar" (215). Rather than legible symbols, they are entangled and makeshift beings. Consider the following lines from the "song" inside the long poem "Al Aaraaf":

> *Ligeia! Ligeia!*
> *My beautiful one!*
> *Whose harshest idea*

Will to melody run,
O! is it thy will
On the breezes to toss?
Or, capriciously still,
Like the lone Albatross,
Incumbent on night
(As she on the air)
To keep watch with delight
On the harmony there? (Poems 109)

The "lone Albatross," inevitably remindful of the sacrificial bird in Coleridge's *The Rime of the Ancient Mariner*, is compared to one "Ligeia," who is a volatile nymph or a vague musical spirit. This kind of second-order comparison is further muddled by the false alternative between air and night. The albatross's poise is oxymoronic, "capriciously still," as if to enhance the arbitrariness of the literary device that ascribes intelligible meaning to an animal. The arbitrary power of words in conferring agency upon animals is highlighted also in the poem "Fairy-land," in which a "filmy" form, or moon, becomes "like—almost any thing— / Or a yellow Albatross" (*Poems* 141). As language turns upon itself, the poetic speaker seems careless of animal individuation, using the animal as an artifice, "subjected to human intent," in the words of Donna Haraway (*The Companion* 28).

Yet, this foregrounding of arbitrary representation can serve as a veiled critique of the transcendentalist view of human intent as being dependent on "higher laws," based on the belief in a necessary, trustworthy correspondence between "words" and "natural facts," as Ralph Waldo Emerson contended in his 1836 "Nature" (20). In "The American Scholar," Emerson additionally asserts that "one nature wrote and the same reads" and "life is our dictionary" (58, 61), reflecting an organic conception of language, transmittable to (and from) the spiritual realm. This was something Poe could not subscribe to, first, because he saw language as a matter of craftmanship and composition; and, second, because he foregrounded the distortion of human affect and emotion in the apprehension of the object. Thus, on the issue of representation, Poe's praise of artifice thwarted Emerson's ideal of nature as well as his conception of language. Moreover, Poe's ascription of speech (although fanciful) to animals other than humans challenged the idea that representative men are the only ones who can relate things and account for them and other beings. As argued by Boggs, the question of whether Poe's fictional gold-bugs or orangutans have semiotic agency or if, in his poetry, the Raven knows the meaning of his utterances, undermines the certainties that "the subject is always verbal and the verbal subject always already human" and leads us "to inquire anew into the relationships between verbal representation and subjectivity with all their ethical and legal repercussions" (115).[6]

The idea that poetic language is taught by the bird's song surfaces in a couple of Poe's poems, most notably in "Romance," where the sentimental fancy is compared to "a painted paroquet," "a most familiar bird" that

> *Taught me my alphabet to say—*
> *To lisp my very earliest word*
> *While in the wild wood I did lie,*
> *A child—with a most knowing eye. (Poems 128)*

Interestingly, in this poem (whose first version, "Introduction," is from 1831 and hence prior to Emerson's "Nature"), the poet seems to toy with the idea of the "eye" whose knowledge comes from lying "in the wild wood"— nonetheless, the alphabet the wide-eyed child acquires is not bequeathed naturally but from a domesticated bird. Standing for romance, the paroquet is a sentimental bird that contrasts with the "tumult" inflicted upon the poet by "eternal Condor years" in the second stanza (128), pointing to a literary tradition where the tamer bird represents love and the raptorial one a challenge to the dominion of humans.[7] The artificiality of the literary device of the talking bird is further underscored because the trustful and loving condition is brought about not by a dove or nightingale but by a "painted paroquet"— or rather a reflection of one (what the poet sees is the bird mirrored upon some "shadowy lake").

We know from "The Philosophy of Composition" that a parrot was a discarded option for "The Raven." Poe's interest in domestic talking birds possibly reflects the dissociation of a "culture of sensibility" that took place in the eighteenth century, simultaneously attached to the sentimental and the mechanical. According to Alex Wetmore, this double-edged inclination was due to the social development of "an affective relationship to mechanical phenomena," including mass copying (134). This would account for the seemingly incompatible view of "sentimental talking birds" as "symbols of (often illegitimate) automatism and as recipients of (often legitimate) sympathy," the latter enhancing "their proximity (whether affective, intellectual or otherwise) to humans" (136). Corresponding to this line of thinking, in "The Raven" the bird's repetitive utterance, its "stock and store," comforts the speaker ("beguiling my sad fancy into smiling"; *Poems* 366) who further conjectures that it was learned out of empathy for the "melancholy burden" of its "unhappy master" (367). The thematization of uttering animals as standing halfway between man and the machine brings an interesting angle to Poe's imagining beyond the human (i.e., posthuman) or outside realms of enlightened reason (i.e., posthumanist). James Berkeley has argued that Poe emphasizes "language's ability to circulate . . . forms that are curiously detachable, even prosthetic, with regard to individual agency" (372), which, seen from our point of view, destabilizes not only language as a means of representation but also animals as representing objects, when they become speaking subjects. Berkeley sees in "Maelzel's Chess-Player" a pre-cybernetic automaton that threatens the "overturning of the humanist liberal subject" through "mimetic debunking" of abilities that account for our exceptionality as a species (357–58); as a talking bird, Poe's Raven functions in a similar manner.

The humanist liberal subject, however, was only a subset of the human range. The possessor of the *logos*—as feminist, postcolonial, and eco-critical thinkers have exposed—was white and male. This begs the question of where to place women on the continuum of accessibility to speech and in relation to their own representation, or (as animals) to their own naming. Poe's poem "To—," first penned in 1829, is another gloss on the topic of the sentimental talking bird. It establishes a parallel between "the wantonest singing birds" and "the lip begotten words" of an unnamed beloved whom T. O. Mabbott presumed to be Elmira Royster (*Poems* 132). The Raven might be a ventriloquist for one "Lenore," the efficacy of whose name is challenged since only the angels seem to know it. We are, in fact, told that the beloved is "nameless *here* for evermore" (emphasis in the original). This type of anonymity is also projected onto the Raven, whose answer to the question "Tell me what thy lordly name is on the Night's Plutonian shore!" is the automatic "Nevermore" (366).

Jacques Derrida has suggested that the mourning emanating from nature derives not from an inability to address us but from its animate beings having "received [their] name" from the (hu)man (19). Thus, we might ask whether the Raven stages a crisis of control related to the masculine subject's entitlement to name or speak. In "The Raven," the speaker is unable to maintain his prerogative over the seemingly otherworldly animal or the dead beloved woman. Moreover, as hinted by Boggs, the suggestion that the Raven might be an articulate double of the lost woman "subverts that way of . . . hierarchizing the body and language" (126) posited by "The Philosophy of Composition." Specifically, the contention that "the death . . . of a beautiful woman is, unquestionably, the most poetical topic in the world," or that the bereaved lover's "lips" are the "best suited" to tell it (*Poe's Critical Theory* 65), is challenged by the prospect that the Raven may embody a different speaker and an alternative story.

Arguably, other animals speak for dead women in Poe's fiction. In "The Black Cat" an animal exposes the unnamed narrator's murder of his wife. In the short story "Ligeia," the eponymous character (whose name is the same as the abovementioned spirit in "Al Aaraaf") with "raven-black" hair (*Tales* 2:312) is the most learned of Poe's women, a poet who dies shortly after composing "The Conqueror Worm" only to then reincarnate in the form of "Rowena." So could it be that as the Raven does for Lenore and the entombed feline does for the murdered woman in "The Black Cat," the conqueror worm speaks for Ligeia? Reified as beautiful dead corpses, Poe's women have also been "com-posted" into animal-human-undead "critters," the focus of our next section.

Undecidable Origins, Composite "Critters," and "Com-Posts"

The Raven has a disturbing ontological status. A "bird of yore," it is surmised to have come "from the night's Plutonian shores" but also to have been sent either by "the angels" or a "Tempter" and to be privy to what exists "in Gilead" (*Poems* 367–69). This is compounded by the anthropomorphic characteristics generally ascribed to this bird species in Poe's time. In the words of Michael Ziser, "Credited even in the nineteenth century with a high degree of intelligence, a tendency to invade human territory, a high level of conjugal fidelity among mated pairs, an ability to mimic the human voice, and a consuming passion for rotting flesh, the raven provided the most uncanny zoosemiotic display of the more-than-human" (26).[8]

The Raven is a liminal creature that merges transcendence and the animal, the spiritual and the sensual, (after)life and death, the ancient and the contemporary, affectionate companionship and impish perversity. Though personified as male ("his soul," "his eyes"), the Raven's gendered performance is ambivalent, "with mien of Lord or Lady," "many a flirt and flutter" (*Poems* 366). Joan Dayan asserts that Poe's "reconstructions depend upon experiences that trade on unspeakable slippages between men and women, humans and animals, life and death" (244), and I would venture that perhaps these slippages are not unspeakable but uttered by sundry bodies buried in "some under-current . . . of meaning" (*Poe's Critical Theory* 70) that at best can only be overheard. Furthermore, the suggestion that the Raven's body could be the reincarnation of a dead woman reminds us of Poe's attraction to mesmerism, also called "animal magnetism" (see Taylor 196–97; Boggs 128). But this sympathy of bodies, which paradoxically entails decomposition, also leads to a rapport between the human's and the animal's body. The Raven is announced by "rapping" when the speaker is "nearly napping" (*Poems* 364), and this hypnagogic state could facilitate a migration to another being, even if not a full reembodiment, as the speaker's soul "from out that shadow . . . shall be lifted nevermore" (367).

It is also possible that the end of "The Raven" anticipates Stéphane Mallarmé's "elocutionary disappearance of the poet" (211), as suggested by the combined reading of his translation of Poe's poem and the accompanying lithographs by Édouard Manet in the art album *Le Corbeau* (1875). In the final plate, what remains of the room is an empty chair and a composite of bird shapes, or a mirrored shadow that blocks the door and extends to the ground, besides an avian figure that ominously occupies the speaker's seat (fig. 2.1).

From the angle of posthumanist animal studies, the speaker that disappears can, in turn, be interpreted as the human white male subject, holder of logocentric knowledge ("forgotten lore"; *Poems* 364)—a reading that prefigures the criticism on Western "enlightened" notions of the human by Rose Braidotti and Cary Wolfe. In the words of the latter, "posthumanism . . . isn't posthuman at all—in the sense of being 'after' our embodiment has been transcended—but is only posthumanist, in the sense that it opposes the fantasies of disembodiment and autonomy, inherited from humanism itself" (Wolfe xv).

In "The Raven," the speaking subject stands under the animal shape, even if the bird's body itself displays human traits and shows signs of having been tamed and altered—"though thy crest be shorn and shaven"—

Figure 2.1.
Édouard Manet, pl. 4 in *Le Corbeau*, translated by Stéphane Mallarmé (Paris: Richard Lesclide, 1875).

by its "unhappy master" (*Poems* 367), attesting to what today we call, after Michel Foucault, "biopower" (184, 189), or the control over differentiated organisms. Moreover, the bird is composed of reified body parts: prosthetic, human, and otherworldly. Athena's head is an example of the first, the poet's heart of the second, and the demonic eyes of the third.

At this point, I venture to link this multipart animal to not only Donna Haraway's use of "critters"—"lively beings" that "are always relationally entangled rather than taxonomically neat" (*When Species Meet* 330n3)—but also her proposal, in *Staying with the Trouble*, of replacing the term "post-humanist" with "compostist" (97). "Com-post" stands for organic litter, or humus, and how we live with (*com*) it and one another in a posthumanist stage that might do without individuated entities and reckon with entangled organisms and assemblages. Against individualism, the exclusive properties of our bodies and our dwellings, and "human exceptionalism" (11), the notion of com-post might illuminate the functions of the Raven and other animate entities in Poe's poetry, but there are caveats to this association. For Haraway, the "articulation of assemblages" (42) should emphasize an intersubjective move whereby, understanding that we stand in relation with other beings, we should cultivate our relatedness to them, as indicated by her ubiquitous expressions "making kin" and "becoming *with*" that embed a will to thrive. Poe's unhappy dénouements hardly fit this project; as suggested later, the eco-critical angle that best becomes them is probably Timothy Morton's "dark ecology." Another problem is that, although Haraway has famously welcomed the cyborg into the "mix of historically situated organic and technological species, both human and not-human" (*Staying* 211n7), her focus is on the richness of Gaia, the humus (to the point of wanting also to replace the term "humanities" with "humusities"), while for Poe mechanical assemblage takes precedence over the organic. Thus, in a well-known passage where he criticizes Coleridge's distinction between fancy and the imagination, composition overrules creation:

> The fancy as nearly creates as the imagination, and neither at all. Novel conceptions are merely unusual combinations. The mind of man can imagine nothing which does not really exist; if it could, it would create not only ideally but substantially, as do the thoughts of God. It may be said, "We imagine a griffin, yet a griffin does not exist." Not the griffin, certainly, but its component parts. It is no more than a collation of known limbs, features, qualities. (Poe, *Essays* 277)

Informed by this passage, the Raven's "philosophy [and physiology] of composition," becomes uncannily similar to the griffin's, and even more a matter of dissectible elements, concurring with a poetics of animal autopsy.

Before Haraway, Jed Rasula used the concept of "compost" in the main title of a curious book on organic intertextuality, in which he characterizes Poe's morbidity as an inability to regard death as a condition for communing with others and to see a poem as a living matter of a community of "leavings."[9] Thus, he disparagingly contrasts Poe with Whitman: "For Poe, the thought of death is such a delectable stimulation that his work is preordained to macabre hallucinations. For Whitman, on the other hand, death is the supreme organic event, the measure of all creaturely striving" (52). However, it might be that Poe's trick is to dramatize the terror of death to the (male) human(ist) speaking consciousness, that is, the dread of not surviving as such nor retaining the capacity for "unnatural" aesthetics—like thematizing the "death of a beautiful woman" (*Critical Theory* 65). Moreover, in Poe's work, suspicious characters and presumed victims often (re)appear as both human and beast: lunatics, dwarves, primates, cats, telling of an ever-living corruption, depicted in "Ulalume" as "cheeks where the worm never dies" (*Poems* 417).

Poe was well read in natural history and especially interested in processes of decomposition and metamorphosis, namely, through worms and bugs, as Susan Elizabeth Sweeney has shown in her article "Insects, Metamorphosis, and Poe's 'The Gold-Bug'" (in *Animals in the American Classics*, a companion to this present volume). Poe's knowledge of ravens, Ziser has suggested, might have been enlarged by *Ornithological Biographies* (1831–39) by John James Audubon, who also wrote about "carrion crows," a common name for the vulture. We are reminded here of the similarity of images between a vulture preying on the poet's heart in "Sonnet—to Science" and the speaker's plea in "The Raven" to "take thy beak [i.e., its instrument of speech] from out my heart" (*Poems* 368). Is the speaker then dead, already a disembodied consciousness uttering his dirge through the bird that feeds on him? Or has he been replaced by a living surrogate? These images suggest that the decomposition process shared with animals is ever more certain than our divine progress, posing a threat not only to the Enlightenment ideals of knowledge and to the notion of scientific advance that underpinned the Industrial Revolution but also to the Romantic image of the divinely inspired poet. Foremost, they tell of how the human will be outlasted by the predator and by the parasite.

In the parasitic ranks, with a sour pandemic aftertaste, we are bound to think of "The Conqueror Worm." Described as a "crawling shape," it is associated with the "vermin fangs" at which "seraphs sob" (*Poems* 325). The poem employs theatrical imagery and foregrounds backstage mechanics, with "vast formless things / that shift the scenery to and fro" and exhibit "Condor wings" (325), thus conforming to Poe's cherished aesthetics of "the ludicrous heightened into the grotesque" (*Collected Letters* 1:85). The latter term, famously used by Poe in the title *Tales of the Grotesque and the Arabesque*, points to an architectural style "created by melding incongruent visual components" (Bryant 3) where animals assume distorted traits and are combined with other forms in singular design.[10] In a poem that makes us think of every animate form as delusional or shape-shifting, of angels dressed up as spectators ("An angel throng, bewinged, bedight / In veils, / and drowned in tears"; *Poems* 325), the conqueror worm is the ultimate composite, at work in the process of decomposing and (with some wishful thinking) composting. It is possible that this worm, "[a] blood-red thing that writhes from out" (326) is also a sign of the excruciating process of metamorphosis, the gory subversion of Emerson's rings of natural forms around the focal, mastering man, as he puts it in the epigraph to "Nature": "And, striving to be man, the worm / Mounts through all the spires of form" (Emerson 5).

Sweeney's analysis of "The Conqueror Worm" in the context of its insertion in the final version of the tale "Ligeia," where the protagonist is compared to "a moth, a butterfly, a chrysalis" (*Tales* 2:314), invites the speculation that "a larva that develops from eggs laid in human remains" (Sweeney 62) most effectively describes the transmogrification of Ligeia into the narrator's second wife, Rowena. There are grounds, then, also for reading gender struggle into the poem, along with animal-(hu)man liminality. The phallic, reddish contortions of the worm add sexual connotations, suggesting man's downfall is tied to his yielding to instinct. The poem's final couplet, "That the play is the tragedy, 'Man,' / And its hero the Conqueror Worm" (*Poems* 326), points to the almost certain destruction of the human, as well as the adherence to organically perishable animal matter rather than to a godlike soul. Indeed, the godly assumptions of "Man," along with his confidence in speech, or the "power of words" to access the divine, are mocked, if we decode the following lines as referring to the biblical account of men created in God's image: "Mimes, in the form of God on high, / Mutter and mumble low, / . . . / Mere puppets they" (325). This hybrid, monstrous, and ungodly figure of "Man" might conform to the "dark ecology" of philosopher Timothy Morton, who supports "embracing the monster" to face "the disgusting real of ecological enmeshment" (124). According to Morton, a dark world vision allows for the acknowledgment of the human sense of interference and othering, along with the realization that interdependence, far from harmoniously symbiotic, is contingent and fundamentally lacking as long as humans act as both disturber and disturbed.

The idea of assembled entities need not be utterly posthuman, even if it seems post-God. To this effect there is a line that Poe uses as an example of an "unintentional instance of a perfect English Hexameter formed upon the model of the Greek" in his "Notes upon English Verse," which Mabbott thinks very probably came from his own pen: "Man is a complex, compound, compost, yet is he God-born," (*Poems* 339). The comma at the end seems important, suggesting a continuation of the sentence in the next line that might end in a question mark, or necessitate further compounds, further entanglements of matter. Morton insists that our exclusionary perception apparatus often precludes such entanglements, reminding us that "[a] monster is something seen by someone (from the Latin *monstrare*, meaning to show). Monstrosity is in the eye of the beholder" (65–66). The necessity of facing "man" as a beastly, writhing, monster-angelic "critter" arguably underlies the staging of "The Conqueror Worm" in a manner whereby readers become spectators, just like "the angel throng," forced to watch through the initial imperative "Lo!" (*Poems* 325).

Another of Poe's poems that convokes the reader, while featuring composite creatures as well as the possible extinction of (the hu)man, is "The Bells." Each of the four sections, corresponding to a different epoch (silver, gold, bronze, iron), starts with an imperative, thereby presenting us with different stages of life and of human history and beyond: "Hear the sledges of the bells" (first section); "Hear the merry wedding bells" (second); "Hear the loud alarum bells" (third); "Hear the tolling of the bells" (fourth). Musical language, the core of poetry for Poe, is foregrounded as performative, mechanical . . . and animal. Like the Raven, the bells not only "tell" but, in their varied performance, are taken as prophetic as they "foretell . . . in a sort of Runic rhyme" (*Poems* 435–36). Moreover, the bells are disturbingly personified as figures that have lost the capacity for speech and are reduced to utterances of terror: "Too much horrified to speak, / They can only shriek, shriek" (*Poems* 436). Their sounds, both machinal ("jingling," "chiming," "twanging," "clanging") and "animal" ("throbbing," "sobbing," "moaning," "groaning"), interpenetrate the elemental beings around them: "What a liquid ditty floats / To the turtle-dove that listens while she gloats / On the moon!" (*Poems* 436). These last lines are suggestive of a communicative interaction between bells, bird, and the earth's satellite.

Therefore, "The Bells," just like "The Raven," can be interpreted in connection with Barbara Johnson's argument that Poe confounds the central Romantic distinction between the mechanical and the natural by using nonhuman language to foreground the "natural passion involved in repetition" (99). By the same token, as argued by Ziser, when Poe stages "the more-than-human dimension of poetic speech," he upsets the boundaries "between the repetitive, mechanistic, empty, 'languages' of animals and the full language of man" (30n23). In "The Bells," as in so many of Poe's poems (e.g., "The Haunted Palace" and "The City in the Sea"), blurring boundaries or becoming a hybrid or a com-post comes as a consequence of the eradication of humanity as we know it. In the fourth stanza, "the people / They that dwell up in the steeple" are revealed to be "neither man nor woman— / neither brute nor human," but "Ghouls" (*Poems* 437). They later oscillate between vampire-like and zombie-like beings; or, in a line discarded from the poem, "pestilential carcasses disported from their souls" (439)—much like "the troop of echoes" that becomes "a hideous throng" in "The Haunted Palace" (316). Along with an allegory on the stages of life and history, "The Bells" is a poem of echoes and of dark ecology, its distortion of language arguably being what is left of organisms once conceived as human. Even if the "King . . . who tolls" (437) is portrayed yelling "a happy Runic rhyme" (438), the rough and harsh sounds thwart the possibility of intersubjective participation in the superficial meaning of the lines. Poe's composite beings cannot cooperate in meaning making or commingle in survival, as they remain shadowed by a figure of authority, generally masculine, that entertains the nostalgia that the symbol might be a portal for an original, hidden, and/or transcendent truth that is ultimately unattainable. In the end, generally one lonesome critter is left that looms over a deserted room ("The Raven"), a devastated stage ("The Conqueror Worm"), or an apocalyptical landscape ("The Bells"), delusional in its mastery, conquest, and kingliness but defeated by the grotesque.

Artistic Renderings of Poe's Animate Com-Posts

Let us now investigate how a few international visual artists have envisioned Poe's poetic animals, starting with Edmund Dulac (1882–1953), who embraced Poe's idea of placing the grotesque (animalized and mechanically stylized caricatures) alongside the arabesque (abstract, coiled lines simulating foliage patterns) in the headplate he created for "The Bells" in 1912. Because a drawing cannot produce "tintinnabulations" (*Poems* 435), the French-born artist, whose "Orientalizing" strokes gained much renown in Edwardian England, depicted a whirlpool of what look like skeins. The arabesques we see in the drawing are probably the bells' ropes, which, endowed with the texture of animal fur, seem enmeshed with the bells and heads, thus conveying the entanglement of the animal and the mechanical inherent to the idea of grotesque. Moreover, the heads are rather animal-like, even simian, or else human gargoyles, and some of their open mouths replace the bell's clappers, arguably alluding to the more-than-human utterances, or the sort of jingled chatter, produced by the poem (fig. 2.2).

Addressing another of Poe's concerns, the theatrics of composition, let us turn to how "The Conqueror Worm" is depicted by František Kupka, a Czech painter and graphic artist who was one of the pioneers of cubism and abstractionism, spending most of his career in France. Produced in 1900, the oil painting *The Conqueror Worm* strikes us with ominous symbols, such as swastikas and skulls, which undergird the stage and frame the "show" (*monstrare*) of a surprising monster: not the worm of Poe's poem but a female figure with large feathery wings (fig. 2.3).

Tony Magistrale and Jessica Slayton have read this figure as an interpretation of Ligeia, the poem's author, turned into a "warrior-survivor devoid of sentimentality or pity," a dominatrix, or at any rate a sort of vampire woman: "It is no accident that the stage upon which she struts is decorated with swastikas, transforming Ligeia . . . into a Valkyrie" (33). Upon closer inspection, it seems that the breasts and womb of this putative Valkyrie

Figure 2.2. Edmund Dulac, headplate for "The Bells," in *The Bells and Other Poems*, 1912.

Figure 2.3. František Kupka, *The Conqueror Worm*, 1900. © František Kupka / ADAGP, Paris / SPA, Lisboa, 2024.

might suggest the body of an insect or, with some imagination, a pupa, or chrysalis in the process of metamorphosis—which connects with Sweeney's reading of the tale "Ligeia." But the figure's outer body opens wings that do not resemble those of a butterfly but rather a raven's, a condor's . . . or a peacock's. As the pied feathers of the latter occur only among males, this might suggest a hieratic figure who is fluid in gender and species. Additionally, the details exude a mystifying reflexivity that recalls the "demon-traps" and "*cock*'s feathers" ascribed to the craft of poetry in "The Philosophy of Composition" (*Poe's Critical Theory* 61).

This animalized figure points to the hardiness of a "com-post" that might outlast, as perhaps Ligeia does, the "tragedy 'Man'" (*Poems* 326), even while it places the woman, sensual and feral, somewhat outside the human-animal species divide. But how would a woman artist illustrate this com-position? Léonor Fini (1908–96), who was born in Buenos Aires and pursued her career in Paris where she would become part of the predominantly male surrealist movement, provides some answers in the more than sixty illustrations she made of Poe's work. Notably, one of her illustrations for "Ligeia" (fig. 2.4), in a 1952 edition of *Contes mystérieux et fantastiques*, features a procession of winged women.

In this artwork, the omnipresent raven looms over the women's bodies, and its wings and tail intersect and com-pose the central figures. The animal silhouette is headless, yet a severed female head appears at the bottom of the image. One of the women is crouching and another seems to be crushed, some of them seem veiled, others ecstatic; they stand for countless women. And we do not know whether they descend from the beastly bird (a parody of women angels?) or if they are meant to reincarnate into it, perhaps as Lenore does in "The Raven," hovering like a *nemesis*, the Greek winged figure of divine retribution, over the speaker of Poe's poem, whom we might associate with the bird's "unhappy master."

Fini's work is loaded with hybrid, prosthetic, skeletal, and spiritual female bodies, arguably as a result of her obsession with transgressive male

Figure 2.4. Léonor Fini, "Ligeia," in Poe, *Contes mystérieux et fantastiques*, 1952, p. 47. © Léonor Fini / ADAGP, Paris / SPA, Lisboa, 2024. Bibliothèque nationale de France.

authors, including Poe and Sade. For the frontispiece of *Contes mystérieux et fantastiques*, she also drew a humorous but disturbing self-portrait of the woman artist as a raven (fig. 2.5). Her claws are strong and resemble tree bark and roots, while, from the hair, another raven and perhaps its fledgling seem to take form.

Despite her ghastly figure, she is poised, indeed extremely *composed*, and offers herself to our gaze—a gaze that wanders over her body and suspects a nascent pregnancy without knowing which critter might be begotten inside such a furry and feathery womb. Fini's portrait offers us a strong revision of Poe's poetical use of animals: their bodies attach to other beings, leaving the viewer to wonder about the painful shape-shifting created by this female psyche. Could this be a woman trying to explode patriarchal conceit and its control over human boundaries?

Throughout this chapter, I have traced how in Poe's verse animals suggest a composite poetics and a mechanical aggregation of parts with disturbing effects. Poe's critters unsettle the correspondence of representation by replacing it with com-position. Furthermore, by exposing the frail construction of man's autonomy, they challenge the idea of who or what can be representative in a constituency, holding the upper hand in matters of life and death or decisions over sentient beings.

Through the utterances of animate composites—Raven upon Pallas, ghoul-played bells, angels who disavow man and "affirm . . . the Worm" (*Poems* 326)—Poe seems to toy with the idea of the "seat of reason" being overturned either through automation or the revenge of other beings or both. I would like to say that these other beings form a viable *com-post*, as in the hopeful neologism created by Haraway, but in Poe the snare of solipsism is stronger, the power of words and of beings is perversely irreconcilable, and there is hardly a way out for the (hu)man who still behaves as a humanist master.

"Man" is the tragedy, with his mental musings and his nightmares about fleshed-out grotesqueries, and his hasty presumptions about speech, rationality and . . . the deaths of beautiful women. In the majority of the images chosen for our analysis, winged animals and women change shape, assuming demonic and prophetic powers that come from their resurgence, their polyvalence, their hardy multiplication, their wombs and eggs. More than threatening to turn against the master, they transgress the division between what is natural and what is constructed or staged, what is specific to a species and what is still to be gathered about how bodies are assembled and how beings are interdependent.

Figure 2.5. Léonor Fini, frontispiece in Poe, *Contes mystérieux et fantastiques*, 1952. © Léonor Fini / ADAGP, Paris / SPA, Lisboa, 2024. Bibliothèque nationale de France.

Notes

This work is financed by Portuguese national funds through the FCT – Foundation for Science and Technology, I.P., within the scope of the projects UIDB/00114/2020 and UIDP/00114/2020. I also acknowledge Emron Esplin and John Gruesser for their linguistic assistance and precious advice on clarification.

    1. See *Ravens—Intelligent Rascals of the Skies*, directed by Heribert Schöller, available at https://www.youtube.com/watch?v=D6s3u0624P8.

2. According to Derrida in *The Animal That Therefore I Am*, the problem of intentionality unfolds in the ability to respond versus merely react, thus linking intention with the use of language, even if speech might also be disputed as a capacity exclusive to humans (8). Derrida will also explore other distinguishing traits, such as, in the wake of Jeremy Bentham, the capacity to suffer and—extensively—the autobiographical genre (see especially 27–35), whereas other authors also flag the sense of humor or the imagination (see Copeland's overview, esp. 96–97).

3. "The self-love and arrogance of man will persist in denying the reflective power to beasts, because the granting it seems to derogate from his own vaunted supremacy" (*Tales* 1:478).

4. Haraway, *When Species Meet* (330n33). For analytical reasons, I here leave plants and fungi aside while including composite ghouls and cyborgs.

5. It is, therefore, difficult to situate them in the cline from zoomorphism (animalization) to anthropomorphism (humanization) presented by Greg Garrard's typology of animal uses in literature. Only the juvenile piece "Oh, Tempora! Oh, Mores!," albeit containing the abovementioned conundrum about "the seat of thought" for man or brute, uses animals for clear belittlement purposes: the comparisons of the poem's target (Bob) to a "fish," or of his sharp look for business to "cat's eyes," or of his conduit to "a proper ass" (*Poems* 11–12), all fall into "crudely anthropomorphic projection[s] of despised human qualities onto these animals" (Garrard 160).

6. For an analysis of the implications of an orangutan in the place of a supposedly foreign-speaking murderer in "The Murders in the Rue Morgue," see Boggs 111–32; for how the humanlike behavior of the orangutan, through imitation and guilt derived from the punishing behavior of his master, destabilizes "the limits of human exceptionalism," see Philip Phillips, "'At the Same Time More and Less than a Man': The Ourang-Outang in Edgar Allan Poe's 'The Murders in the Rue Morgue'" (42–46).

7. See Brycchan Carey, Sayre Greenfield, and Anne Milne in the "Introduction" to their edited volume, *Birds in Eighteenth-Century Literature* (7).

8. The misanthropic and gothic undertones are also well-known. In *Dictionnaire des symboles*, Chevalier and Gheerbrant remark that it was Romanticism that conferred upon the raven a status of ill omen; in several traditions, it has meant salvation or resurrection, and the alchemists have long valued the "tête du corbeau," the name given to the black matter, which must be imbibed, sublimated by fire (285–86).

9. Rasula's *This Compost* is based on the idea of intertextual or literary "leavings" that contribute to living matters in poems.

10. Note that in "The Murders in the Rue Morgue" the violent annihilation of two

spinsters, later revealed to be perpetrated by an orangutan, is called "a *grotesquerie* in horror absolutely alien from humanity" (*Tales* 1:588; emphasis in the original), yet the detective and the narrator inhabit a "grotesque mansion" (530).

Works Cited

Berkeley, James. "Post-human Mimesis and the Debunked Machine: Reading Environmental Appropriation in Poe's 'Maelzel's Chess-Player' and 'The Man That Was Used Up.'" *Comparative Literature Studies* 41, no. 3 (2004): 356–76.

Boggs, Colleen Glenney. *Animalia Americana: Animal Representations and Biopolitical Subjectivity*. New York: Columbia University Press, 2013.

Braidotti, Rose. *The Posthuman*. Cambridge, UK: Polity, 2013.

Bryant, Clinton M. "An Effect All Together Unexpected: The Grotesque in Edgar Allan Poe's Fiction." University of Vermont, Graduate College Dissertations and Theses, 2017. https://scholarworks.uvm.edu/graddis/705.

Carey, Brycchan, Sayre Greenfield, and Anne Milne, eds. *Birds in Eighteenth-Century Literature: Reason, Emotion, and Ornithology (1700–1840)*. Cham, Switzerland: Palgrave MacMillan, 2020.

Chevalier, Jean, and Alain Gheerbrant. *Dictionnaire des symboles: Mythes*, rêves, *coutumes, gestes, formes, figures, couleurs, nombres*. 2nd ed. Paris: Robert Laffont, 1982.

Copeland, Marion W. "Literary Animal Studies in 2012: Where We Are, Where We Are Going." *Anthrozoös* 25, supp. 1 (2012): 91–105. DOI:10.2752/1753037 12X13353430377093.

Dayan, Joan. "Amorous Bondage: Poe, Ladies, and Slaves." *American Literature* 66, no. 2 (1994): 239–73.

Derrida, Jacques. *The Animal That Therefore I Am*. Edited by Marie-Luise Mallet. Translated by David Wills. New York: Fordham University Press, 2008.

Emerson, Ralph Waldo. *Essays and Lectures*. New York: Library of America, 1983.

Foucault, Michel. *La volonté de savoir*. Paris: Gallimard, 1976.

Garrard, Greg. *Ecocriticism*. 2nd ed. London: Routledge, 2011.

Gruesser, John Cullen, ed. *Animals in the American Classics: How Natural History Inspired Great Fiction*. College Station: Texas A&M University Press, 2022.

Haraway, Donna. *The Companion Species Manifesto: Dogs, People, and Significant Otherness*. Chicago: Prickly Paradigm Press, 2003.

———. *Staying with the Trouble: Making Kin in the Chthulucene*. Durham, NC: Duke University Press, 2016.

———. *When Species Meet*. Minneapolis: University of Minnesota Press, 2008.

Johnson, Barbara. "Strange Fits: Poe and Wordsworth on the Nature of Poetic Language." In *A World of Difference*, by Barbara Johnson, 89–99. Baltimore: Johns Hopkins University Press, 1987.

Kress, Gunther, and Theo van Leewen. *Reading Images: The Grammar of Visual Design*. 3rd ed. New York: Routledge, 2020.

Magistrale, Tony, and Jessica Slayton. *The Great Illustrators of Edgar Allan Poe*. New York: Anthem Press, 2021.

Mallarmé, Stéphane. *Oeuvres complètes*. Vol. 1. Edited by Bertrand Marchal. Paris: Gallimard, 1998.

Morton, Timothy. *The Ecological Thought*. Cambridge, MA: Harvard University Press, 2010.

Phillips, Philip E. "'At the Same Time More and Less than a Man': The Ourang-Outang in Edgar Allan Poe's 'The Murders in the Rue Morgue.'" In *Animals in the American Classics*, edited by John Cullen Gruesser, 31–56. College Station: Texas A&M University Press, 2022.

Poe, Edgar Allan. *The Collected Letters of Edgar Allan Poe*. 2 vols. Edited by John Ward Ostrom. Revised by Burton R. Pollin and Jeffrey A. Savoye. New York: Gordian Press, 2008.

———. *Essays and Reviews*. Edited by G. R. Thompson. New York: Library of America, 1984.

———. *Poems*. In *The Collected Works of Edgar Allan Poe*, vol. 1, edited by T. O. Mabbott. Cambridge, MA: The Belknap Press of Harvard University Press, 1969.

———. *Poe's Critical Theory: The Major Documents*. Edited by Stuart Levine and Susan F. Levine. Champaign: University of Illinois Press, 2009.

———. *Tales*. In *The Collected Works of Edgar Allan Poe*, vols. 1 and 2, edited by T. O. Mabbott, with the assistance of Eleanor D. Kewer and Maureen C. Mabbott. Cambridge, MA: The Belknap Press of Harvard University Press, 1978.

Rasula, Jed. *This Compost: Ecological Imperatives in American Poetry*. Athens: University of Georgia Press, 2002.

Sweeney, Elizabeth. "Insects, Metamorphosis, and Poe's 'The Gold-Bug.'" In *Animals in the American Classics*, edited by John Cullen Gruesser, 57–86. College Station: Texas A&M University Press, 2022.

Taylor, Matthew A. "Edgar Allan Poe's (Meta) physics: A Pre-history of the Post-human." *Nineteenth-Century Literature* 62, no. 2 (2007): 193–221.

Wetmore, Alex. "'No Parrot, Either in Morality or Sentiment': Talking Birds and Mechanical Copying in the Age of Sensibility." In *Birds in Eighteenth-Century Lit-*

erature, edited by Brycchan Carey, Sayre Greenfield, and Anne Milne, 131–50. Cham, Switzerland: Palgrave Macmillan, 2020.

Wolfe, Cary. *What Is Posthumanism?* Minneapolis: University of Minnesota Press, 2009.

Ziser, Michael. "Animal Mirrors: Poe, Lacan, Von Uexküll, and Audubon in the Zoosemiosphere." *Angelaki: Journal of Theoretical Humanities* 12, no. 3 (2007): 11–33.

CHAPTER 3

"Animality" and the "Clef of the Universes" in the Poetry of Walt Whitman

AARON M. MOE

Very few people living today, I wager—who want to learn more about what an animal actually is—would turn to a nineteenth-century poet. It seems that the interdisciplinary fields within animal studies would offer the most current knowledge concerning the amazing capabilities of nonhuman animals. Even if someone is not interested in the more field-specific work of Jane Goodall and Frans de Waal (primates), or Cynthia Moss (elephants), or Hal Whitehead (sperm whales), or G. A. Bradshaw (elephants and yet carnivores), or in the theories of Donna Haraway or Jacques Derrida, or in the breakthroughs of affective neuroscience that trace a deep commonality of emotion across the mammalian brain (see Panksepp and Biven)—it is virtually impossible to miss some bewildering feat of an ant, or a cetacean, or a pachyderm, or a bee, or a spider circulating in online spaces. From the peacock spider's mating ritual, to the crow who concocts a sledding scheme, to the fractal dimensions of the clicks within clicks of sperm whale communication, it seems that present-day sources prove most fruitful for exploring and discovering the innumerable facets of the animal.

And yet, the life's work of Walt Whitman ought to cause people to rethink this assumption.

His poetry becomes a massive terrain to be explored, ever inviting the journey of a reader. Focusing on Whitman's pervasive animal imagery is one way to explore this terrain. Though such a study is most definitely warranted, the aim of this chapter goes in a different (though not unrelated) direction, for Whitman infuses the terrain of his poetry and poetics with a philosophy of animality.[1] In what I see to be one of the most daring and tantalizing moves, Whitman suggests that his life's work—stretching across the decades—concerns not necessarily grass and plants and leaves—but rather *animals*. Buried in the essay "A Backward Glance o'er Travel'd Roads" written near the end of his life, he confesses that *Leaves of Grass* is "avowedly the song of Sex and Amativeness, and even Animality—though meanings that do not usually go along with those words are behind all, and will duly emerge" (*LG* 1891–92). Tantalizing?—yes, for he does not expound in the essay what "meanings" he has in mind or exactly how they will "duly emerge" (though he provides a clue). Daring?—yes, for he suggests some sort of relationship between "Sex and Amativeness, and even Animality" that seems to be even more important than the plant, the leaves of grass, so very much at the heart of his poetic vision.

As this daring and tantalizing paragraph unfolds, Whitman argues that sex, amativeness, and animality are "intentionally palpable in a few lines" and that they give the "breath of life" to his "whole scheme." Moreover, he sees the "vitality of it"—that is, the vitality of the "fact of sexuality" along with amativeness and animality—to be found in its "relations, bearings, significance—like the clef of a symphony" (*LG* 1891–92). This chapter, then, focuses on the meanings that emerge when we make Whitman's theory of the clef a starting point for understanding animals and animality within his writings, for the clef is the strongest clue Whitman provides about how the meaning of animality will "duly emerge." As the meaning of animality emerges, I argue that Whitman's insights are just as relevant to us living today—and even to people who will read Whitman's work one thousand years in the future—as they were for the nineteenth-century reader.

Questions include the following: What is an animal? When Whitman witnessed a bird whistling, a spider launching forth silk, flocks of geese filling autumn air with tumult and song, eagles mating, or a dragonfly alighting on a leaf of grass, how did he perceive such creatures and such moments? How should we, as readers, perceive the innumerable representations of animals that scurry, hum, flit, fin, soar, and lumber through the phrases and clauses, the lines and stanzas, of Whitman's poetry? What happens when our context for the animal is nothing short of Deep Space and Deep Time?

To explore these questions, this chapter integrates two ideas from contemporary discourse: hyperobjects and biosemiotics. Whitman's work suggests that any animal—and the cells and atoms of that animal—is nothing less than a fraction of a fraction of a whispering energy "massively distributed in time and space relative to humans" (Morton 1). My not-so-subtle nod toward Timothy Morton's definition of hyperobjects will take the arc of the chapter to develop. Suffice it to say here that Whitman's vision of animality anticipates Morton's theory as the vitality of animality stretches across Deep Space and Time, and we humans, locked in our temporal and spatial bodies, can experience an encounter with only a fraction of such energy at any given moment—an encounter that, due to its ephemeral nature, can be seen as a mere brush.[2] Likewise, Whitman's metaphor of the whispering energy at work within an atom, a cell, a spider, a star, and a soul anticipates some of the ideas emerging from the field of biosemiotics. Again, as it will for hyperobjects, it will take the arc of this chapter to make the meaning clear. Suffice it to say here that the semiotic/interpretive agency at work within the cell is a kind of Whitmanesque *whispering*. For Whitman, this whispering has its origins in the atom, and it forms the basis and foundation for the idea that animality is something massively distributed across Deep Space and Time. Because of Whitman's radical and profound understanding of animality—manifest in both "A Noiseless Patient Spider" and "The Dalliance of the Eagles"—his work is most relevant to discussions today, especially if we seek a philosophical principle to guide our understanding of and interactions with these beings we call *animals*.

The Clef of Deep Space and Time

Whitman first uses the term "clef" in the title "Clef Poem" from the 1856 publication of *Leaves of Grass*. Shortly after the title, he speaks of the "clef of the universes" (*LG* 1856). In later editions of *Leaves of Grass*, the phrase "clef of the universes" disappears, then reappears, finding its final home in the 1891–92 edition within the poem "At the Beach at Night Alone." The fact, though, that Whitman uses the word "clef" in 1856 and then returns to the concept of the clef in both "At the Beach" and "A Backward Glance o'er Travel'd Roads" published well over three decades later suggests that it, indeed, is a weight-bearing term that buttresses Whitman's poetic vision.

Several lines from the 1855 *Leaves of Grass* hint at what Whitman may mean by the phrase "clef of the universes"—lines that point toward the context of Deep Space and Time:

I believe a leaf of grass is no less than the journeywork of the stars,
And the pismire is equally perfect, and a grain of sand, and the
   egg of the wren,
And the tree-toad is a chef-d'oeuvre for the highest,
And the running blackberry would adorn the parlors of heaven,
And the narrowest hinge in my hand puts to scorn all machinery,
And the cow crunching with depress'd head surpasses any statue,
And a mouse is miracle enough to stagger sextillions of infidels.

Each animal, each "grain of sand," and every "leaf"—that is, every part that is also its own whole—ought to be seen in the context of the vast distances and vast time it took for matter to become *animal*, or *plant*, or *element*. The "journeywork" necessary to bring the 4.5-billion-year-old planet into existence in the first place, and then the "journeywork" necessary to bring forth life on the planet can—and ought to—boggle the mind. We, as humans, struggle to imagine Deep Space and Time, as the magnitude of scale is just too much for us to comprehend, but Whitman's idea of the "clef of the universes" helps guide us. We can begin to see the "journeywork of the stars" as a score of music, traveling its long, long arc across Deep Space and Time—an arc that has brought forth this thing we call the animal.

Whitman is interested in the "perpetual transfers and promotions" of matter across multiple forms as well as the "original energy" that infuses the process (*LG* 1855, 1860–61). Such thinking may help us appreciate that flowers did not appear on the earth until 130 million years ago. And yet, prior to the blossom of a plant, matter blossomed into the crystallization of sediments aptly named "Desert Rose." Similarly, we see dendritic, branching patterns in trees, in some snow crystals, in ferns, and yet also in the feathers of birds. Such observations provide evidence that Whitman rightly seeks an "original energy" that is part of the "journeywork of the stars" traveling across the vast distances of Space and Time in the universe.

We see another iteration of journeywork later in the 1855 *Leaves of Grass* when Whitman argues for the divinity and sacredness of the slave being auctioned: "For him the globe lay preparing quintillions of years without one animal or plant, / For him the revolving cycles truly and steadily rolled." Whitman places the human—in this case, the auctioned slave—in the context of a Deep Time that reaches back to when Earth had no life. Whitman adds a few more zeros to the actual age of the planet (4.5 billion years compared with "quintillions of years," which is 1×10^{18}), but he is not altogether *wrong*. He understood the vastness of space and time, and just how long and how far it would take matter and energy to travel in order to bring forth just "one animal or plant"—to say nothing of the journeywork driving forth the innumerable forms of evolved life on planet Earth. For Whitman, both the animal and the process of becoming animal make the most sense when seen as a part within the clef of Deep Space and Time—the clef, that is, "of the universes."

When we perceive an individual animal within the "clef of the universes," we can begin to glimpse how an animal (and a plant and an element) is like one solitary but profound note that has emerged from the long arc of the "journeywork of the stars." Timothy Morton's notion of hyperobjects applies. For Morton, a hyperobject is something so "massively distributed" across space and time "relative to humans" that we strain to even glimpse its magnitude (1). At best, we can encounter just a fraction, or a "slice" as he calls it, of the hyperobject (74). I see such an encounter as a "brush" encounter, a momentary encounter, a fleeting encounter, for Morton suggests that the hyperobject will always recede. It's too massively distributed for humans to ever experience the full object in all of its wholeness. To use Whitman's musical metaphor, we can hold a note—maybe one hundred notes, or one thousand—but the full score of the "journeywork of the stars," of "original energy," is too "massively distributed" across the cosmos for us to ever gain a full grasp of it.

The Plant, the Animal, and the Poem

So, when Whitman nudges readers to "Stop this day and night with me and you shall possess the origin of all poems"; when he claims that his project focuses on "Nature, without check, with original energy"; and when he suggests that a "leaf of grass" of the earth and the *Leaves of Grass* of his poems are both the "journeywork of the stars"—he establishes a deep connection between the poem, the plant, and the animal (*LG* 1855, 1860–61). One cannot understand Whitman's idea of "Animality" without recognizing the kinship between the forms of the plant, the animal, and the poem. In Emerson's essay "The Poet" published in 1844—an essay Whitman read, studied, and responded to with his 1855 *Leaves of Grass*—Emerson points toward a potential origin of the poem, and of the plant, and of the animal. He writes,

"For it is not metres, but a metre-making argument that makes a poem—a thought so passionate and alive that like the spirit of a plant or an animal it has an architecture of its own, and adorns nature with a new thing" (290). One can imagine Whitman reading this statement, pondering it, reflecting on it, searching for the "original energy" that drives forth form—the *spirit* that brings forth the form of a branching tree, or a branching skeletal structure, or the spine of a blade of grass, or the arc of the long, wavering grasslike lines of his stanzas.

In "Whitman Making Books/Books Making Whitman," Ed Folsom establishes how both the plant and the animal infuse Whitman's poetic vision. First, Folsom shows that Whitman hand-stamped the letters of the 1855 *Leaves of Grass* so that they look like vines and vegetation as if to say that the spirit, the energy, bringing forth the vegetation of the earth is also at work to bring forth the form of the poem. However, later in the essay, Folsom shows that Whitman draws the letters of the title page to the 1860 edition of *Leaves of Grass* so they resemble sperm. The "spermatic imagery," of course, resonates not with the plant but with the animal (Folsom). Transforming the letters of his title into vegetation, then into sperm, becomes a not-so-subtle homage to Emerson's idea that the spirit, the energy, bringing forth the architecture of a poem shares a kinship with the energy at work bringing forth the forms of plants and animals.[3]

In *Song of Myself*, Whitman encapsulates the spirit/energy driving the form of the plant, the animal, and the poem all within the metaphor of the *whisper*. It is the "buzz'd whispers" of his "own breath," of "love-root," of "silk-thread," of "crotch and vine" as well as of the whispers just prior to the "sound of the belch'd words" of his "voice loos'd to the eddies of the wind" (*LG* 1891–92)—it is these "buzz'd whispers" at work within the seed, the zygote, and the sounds within syllables that all reinforce Whitman's ongoing homage to and exploration of Emerson's insightful claim.

And it is impossible to trace Whitman's arc of animality without showing its inextricable entanglement with the plant and with the poem. In a stroke of genius, Whitman locates this energy source in something prior to the zygote. Prior to eggs. Prior to sperm. Prior to seeds. Prior to seedless plants—for *Song of Myself* begins with the assertion that "every atom belonging to me as good belongs to you" and that his "tongue" and "every atom" of his "blood" have been "form'd from this soil, this air" (*LG* 1891–92). His quest for the whispers of "original energy"—a term he first uses in the 1860–61 poem "Proto-Leaf" and later moves to the beginning of what becomes *Song of Myself*—leads him to a time prior to the cell as he intuits that it must be found within the stirring energy of the atom. Even though he had only a nineteenth-century understanding of the atom, he is not wrong.[4] He infuses his understanding of the part and the whole (the atom and the animal body) with a search for the "original energy" at work within the atom, the cell, the seed, the zygote, and the sounds of a poem to bring forth form.

Biosemiotics, Abduction, and the Whispers of Eros

In later sections, I discuss Whitman's spiders and eagles, that is, the "whole" of a single organism. However, Whitman revels in part/whole dynamics. He is not only interested in the animal as one of the notes in the "clef of the universes"; he is also interested in energy at work within the parts of the animal. In Whitman's work, I argue that "animality" can be seen as the whispering energy that brings forth the animal (and, perhaps, the plant as well as the poem). Whitman does not limit the energy of "buzz'd whispers" to the plant, nor to the animal, nor to the cells and atoms that make up such bodies; rather, such a vision encompasses grander magnitudes. Near the end of *Song of Myself*, Whitman locates whispering energy beyond the earth: "I hear you whispering there O stars of heaven, / O suns" (*LG* 1891–92). The "whisper" becomes Whitman's weight-bearing metaphor for the energy at work within the seed, the zygote, and yet also the energy of the stars. The whisper is nothing less than *journeywork* that occurs in the "now" of a star, but also the "now" of the trillions of cells sustaining the life of an animal body and the trillions of atoms that make up each cell.

At this point, some readers may dismiss Whitman because his metaphor of the whisper is intuitively based. It seems to lack any sort of grounding in what may actually be true concerning the relationship between atoms, cells, plants, animals, on the one hand, and poems, on the other. However, this is why the emerging field of biosemiotics is so instructive. Over the last couple of decades, the interdisciplinary field has offered a grounded understanding of the agency and proto-consciousness at work on the level of the cell. I draw on two crucial thinkers: the biologist and philosopher Jesper Hoffmeyer and the scientist and theorist Wendy Wheeler. Hoffmeyer establishes the fact that DNA is an inert molecule. It has no agency. The agency of a zygote or a seed to become an animal or a plant takes place at the cell's membranes. "Living cells, through their membranes," he observes, "use DNA to construct the organism, not vice versa." He continues, "It is . . . in the semiotic functioning of the cellular membranes that we shall seek what can be called life's *agency*, its inherent future-directedness, its survival project" (32).

Wheeler sees the agency of the cell similarly. Semiosis—the process of reading signs, interpreting signs, and acting upon signs—is a fundamental principle of life from "the cell all the way up" (270). In order to understand this idea further, Wheeler turns to the concept of abduction. Unlike deductive and inductive inferences, abduction can lead to a "creative thought" that "introduces a new idea." Abduction involves the "strange—obscure and dark—semiotic process whereby signs are read, and interpreted, often without ever necessarily having reached consciousness at all" (273). Though some people may still hold the erroneous view of a cell as a top-down system where DNA dictates and directs the cell's becoming, Wheeler reminds readers (as does Hoffmeyer) that DNA "is simply inert." She continues, "As with a never read book, on its own, DNA cannot cause anything to happen. It is only when read off by proteins in the cell (which interpret its potential meanings in different ways according to time, to higher-level processes and purposes, and to environmental inputs both internal and external to the organism) that its potential meanings (in the building activities of proteins) become activated" (275). Such a perspective invites us to see each cell of every animal (and plant) body to be actively reading, interpreting, and becoming through this intuitive, abductive way of knowing. I am drawn to how Wheeler uses the language of reading and writing to capture the dynamic activity of a cell's becoming.

These "buzz'd whispers" making meaning within the cell can be seen as a facet of Whitman's "Animality." In *Whitman Illuminated: Song of Myself*, Allen Crawford insightfully grounds one of his illuminations in cellular mitosis, or rather, in the abductive, whispering energies at work when a cell splits, that is, when a cell duplicates the genetic code of the DNA from within the chromosomes and then splits into two cells. Crawford's chromosomal strands resemble vines anchored by what looks like the helicopter seeds of a maple tree. Crawford represents the nucleus of the cell as a flower, but, upon reflection, both nuclei simultaneously conjure flowery nipples. The plant, the mammalian body, and cellular mitosis all are momentary manifes-

54 | CHAPTER 3

tations of the "original energy" at work across the cosmos—but also at work within every cell of the animal body—and Crawford's illumination demonstrates the fact that it is impossible, within Whitman's poetic vision, to disentangle the "original energy" at work within and throughout the plant, the animal, and the poem (fig. 3.1). The vitality—that is, the *animality*—at work across the cosmos, then, is something that includes the plant, the animal, and the poem; the journeywork of energy, then, points toward the "buzz'd whispers" at the very origin of the poiesis of all forms throughout the cosmos from the cell, and even the atomic level, *all the way up*.

I suggest, though, that Whitman outpaces biosemiotics. Biosemiotics demands a text. A sign. DNA. Something to be interpreted. Atoms and molecules, though, are governed not by a text but by forces. The leap from matter to the text of DNA and onward toward the animal body occurs out of the erotically charged "buzz'd whispers" of the stars—or rather, the powerful cosmological forces of attraction and repulsion, of desire, of Eros, at work within every atom of the cosmos. Again, Crawford grasps this fundamental

Figure 3.1. An illumination by Allen Crawford. *Whitman Illuminated: Song of Myself* (Portland, OR: Tin House Books, 2014). Reproduced with permission from Allen Crawford and Tin House Books.

POETRY OF WALT WHITMAN | 55

Figure 3.2. An illumination by Allen Crawford. *Whitman Illuminated: Song of Myself* (Portland, OR: Tin House Books, 2014). Reproduced with permission from Allen Crawford and Tin House books.

insight of Whitman. The very forms of the animal, of the process of becoming animal, of animality—have their origins in the elemental forms of water, earth, fire, and air. Crawford demonstrates this Whitmanian principle of the relationship between form and original energy, for he takes Whitman's epic 1,079-word sentence from the 1855 edition of *Leaves of Grass*—which ends with "I tread day and night such roads"—and illustrates it across nine pages. The content of the sentence catalogs a whole host of critters and humans Whitman encounters and paths and roads on which Whitman journeys, but what strikes me is Crawford's decision to subtly morph the initial alligator image into and out of the scaly hills and mountains of the background (fig. 3.2). Are we looking at the form of an animal? Are we looking at the form of a mountain? What is the relationship between the form of a mountain and the form of a reptile? Is it true that both the mountain and the reptile have a backbone? Can we say that the form of the mountain is, too, a manifestation of a kind of animality?

In my effort to explore such questions, I have been moved by the fact that the physicist Roger Barlow uses the metaphor of the "swarm of bees or birds" to most accurately describe the movement of electrons within the atom. If electrons have a kind of proto-animal agency to swarm, to undergo a murmuration like a flock of starlings but within the vast horizon of an atom, then we most assuredly have the justification to ponder some sort of deep kinship between the larger form of a mountain, which includes a near-infinite array of murmurating electrons, and the smaller form of an alligator, which also includes a near-infinite array of murmurating electrons. Both forms manifest branching, bifurcating patterns. The alligator has arteries, veins, capillaries; the mountain has streamlets, tributaries, rivers. The "bones" of both give shape to the outer form. Both exhibit similar patterns of roughness—scales or boulders—all along their respective surfaces.

Whitman seems to have understood that the atom held the "original energy" he so consistently sought, and along with the idea of whispering energies, Whitman gravitates to the energy of the "murmur"—a word he uses fifteen times in the 1891–92 edition of *Leaves of Grass*. "Murmuration" and "murmur" come from the Latin *murmur*, meaning "a hum, a rushing"; and from the Sanskrit *murmurah*, meaning "crackling fire"; and from the Greek *mormyrein*, meaning "to roar, boil" ("Murmur"). The etymology speaks to a profound "original energy" within a murmur. Whitman evokes the *murmur* when he says, "Only the lull I like, the hum of your valved voice" (*LG* 1855). The hums. The whispers. The pre-linguistic boil of sound. The "barbaric yawp" prior to language (*LG* 1855). This whispering murmuration has its origin, Whitman intuits, within the atom. In the late poem "Whispers of Heavenly Death," the erotic energy of such murmured whispers emerges: "Whispers of heavenly death murmur'd I hear, / Labial gossip of night, sibilant chorals"—"labial" pointing both toward the lips of a human that make articulation possible and the lips of a woman's vulva and their "sibilant" song (*LG* 1891–92). But Whitman's point is that such erotically charged, labial murmurs are not limited to the woman, to the human, or to the animal—for the poem continues. It moves toward the whispers in "mystical breezes," and in the "ripples of unseen rivers," and in the "tides of a current flowing, forever flowing," and in the "great cloud-masses . . . silently swelling and mixing" that are akin to some "soul" who "is passing over" (*LG* 1891–92). When Whitman cloud-gazes, he sees the whispering energy of "Nature without check with original energy" as he states at the beginning of

Song of Myself (*LG* 1881–82)—that is, a vitality that shares a deep kinship across all manifestations of labial whispers.

For Whitman, the animal is possible only because of the journeywork of original energy; and original energy is best understood, I argue, as the "buzz'd whispers" of animality within atoms and cells bringing forth innumerable forms of the animals of the planet—all of which are holy. So, when Whitman writes in *Song of Myself* that he and other animals share "the same old law"; when he speaks of the "wing'd purposes" of birds; when he conflates his "barbaric yawp" from the end of *Song of Myself* with the "*Ya-honk*" of the geese near the beginning; when he sings of the erotically charged, electric body of the human in "I Sing the Body Electric"; when he sings of the erotically charged mating eagles in "The Dalliance of the Eagles"; and when he explores the profound forces of Eros, of desire, of hunger, of Sex, of Amativeness—he is, I argue, celebrating solitary notes of animality that make the most sense when seen from within the context of a journeywork across Deep Space and Time, across, that is, the "clef of the universes" (*LG* 1891–92).

The Spider and the Poet's Spiderlike Soul

The moment we take seriously Whitman's well-known "A Noiseless Patient Spider," we deepen our understanding of the "buzz'd whispers" of Whitman's animality. In 1865, this poem was the third of the poem's five sections, previously discussed, titled "Whispers of Heavenly Death" and published in *The Broadway, a London Magazine*. When "A Noiseless Patient Spider" migrated into the 1891–92 *Leaves of Grass*, it became its own poem. This history, however, suggests a rich connection with the whispering murmurs of the "labial gossip of night" discussed earlier—that is, the whispers of the lips, breezes, rivers, tides, and cloud masses that "Whispers of Heavenly Death" explores—and the whispering work of a spider making silk and of a poet's soul making a poem (fig. 3.3). All are manifestations, I argue, of Whitman's notion of animality. All are driven by "the same old law" of Eros, of "Urge and urge and urge, / Always the procreant urge of the world"—the urge of the "buzz'd whispers" of animality, of original energy (*LG* 1891–92). The spider, in a physical sense. The soul, in a metaphysical sense—but still through the very physical sounds of this thing we call language. On a foundational level, the spider spinning silk is a metaphor for the soul spinning

Figure 3.3. A marbled orb weaver with its massive abdomen. Orb weavers are part of the genus *Araneus*, and their spiderlings are capable of ballooning, that is, sending forth strands of spider silk to catch and float through a current of air—not unlike the spider as described in Whitman's poem. Image by Emvats, an Adobe Stock Contributor.

language, and vice versa. Both the spider and the soul have within them the abductive, obscure, stirrings to echo ideas from biosemiotics. The electrons of atoms are awhirr within the spider's body, all making the strands of spider silk to be launched forth; likewise, innumerable sounds are awhirr within the whispers of the poet's soul, all making syllables, soon to be launched forth to the *eddies of the wind*:

> *A noiseless patient spider,*
> *I mark'd where on a little promontory it stood isolated,*
> *Mark'd how to explore the vacant vast surrounding,*
> *It launch'd forth filament, filament, filament, out of itself,*
> *Ever unreeling them, ever tirelessly speeding them.*
>
> *And you O my soul where you stand,*
> *Surrounded, detached, in measureless oceans of space,*
> *Ceaselessly musing, venturing, throwing, seeking the spheres to connect them,*
> *Till the bridge you will need be form'd, till the ductile anchor hold,*

Till the gossamer thread you fling catch somewhere, O my soul. (LG 1891–92)

Undoubtedly, the poem demonstrates how the poet's words become a "gossamer thread" launched forth from the poet—words that can become a connective "bridge" to help an isolated self make some sort of connection and kinship with another being on the planet, spider or otherwise. We see the energy of the "buzz'd whispers" of language in the poem's fourth line as Whitman infuses it with a string of onomatopoeic sounds: *It launched forth filament, filament, filament out of itself.* The *f* sounds crescendo until they coalesce in the final *itself*, generating the auditory rush of silk through the air—*fffffffff*—not unlike the sounds being launched forth into the blank spaces of the poetic page, or, when read aloud, into the open air environing a speaking body. Whitman's language becomes as gossamer as spider silk. It is fragile, but as spider silk is the strongest material on the planet by weight, the language assumes a "ductile" strength as well. After all, the poem's form hinges on two powerful sentences, each arcing across a full five-line stanza.

However, if we foreground how the spider serves as a metaphor for the human soul, we eclipse how the poem helps us attend to the actual, literal spider. The spider, and the human, follow the *same old law* of the whispers of original energy. Some may see the human soul as a more dignified expression of the original energy of the cosmos, but for Whitman, the makings of a spider, too, are a manifestation of original energy. Moreover, his onomatopoeic line helps us perceive what we cannot hear, namely, the sounds of electrons within a spider's silk-producing abdominal glands, all at work in the process of *poiesis*, the making of spider silk. We cannot perceive, with our senses, such sounds. They are beyond our auditory capacity. Whitman, though, in a leap of imagination, helps us realize that such a process has to make a sound. Moreover, the whispering, "noiseless" sounds at work within the spider share a deep kinship with the whispering, noiseless stirrings of human consciousness at work within the poiesis of sound, syllable, word, line, and poem. Labial gossip indeed.

Which is to say, Emerson is right: "In the order of genesis the thought is prior to the form" (290). The *spirit* of the plant or of the animal gives rise to the *form* of the plant or of the animal. Whitman takes, I argue, the Emersonian suggestion of "spirit" and develops it further, naming it as the "buzz'd whispers" of the "original energy" he set forth to explore and discover across the long arc of his life's work. I do not think one can understand what meanings Whitman had in mind that would "duly emerge" from the word "Animality" without exploring the relationship between the animal, the soul of a poet, and original energy. I hazard to say that even though an atheist and a theist would disagree about the *how* of animality—and even though both would use different terms to describe the interiority of a human (consciousness versus soul)—both, I hope, could agree that the energy that brings forth the forms of language shares a deep kinship with the energy that brings forth the form of a web made from the filaments of a spider.

Eagles on the Clef

The energy within a spider's abdomen and that within a poet's soul share a kinship, too, with the energy shared between two eagles as demonstrated in "The Dalliance of the Eagles":

> SKIRTING *the river road, (my forenoon walk, my rest,)*
> *Skyward in air a sudden muffled sound, the dalliance of the eagles,*
> *The rushing amorous contact high in space together,*
> *The clinching interlocking claws, a living, fierce, gyrating wheel,*
> *Four beating wings, two beaks, a swirling mass tight grappling,*
> *In tumbling turning clustering loops, straight downward falling,*
> *Till o'er the river pois'd, the twain yet one, a moment's lull,*
> *A motionless still balance in the air, then parting, talons loosing,*
> *Upward again on slow-firm pinions slanting, their separate diverse*
>   *flight,*
> *She hers, he his, pursuing. (LG 1891–92)*

Masterly crafted, this ten-line, single-stanza poem is one long sentence. It begins with a participle adjective, "Skirting," that describes, profoundly, the implied speaker who, grammatically, is absent from the poem. The poem unfolds as a series of nouns (boldface below). (Most of the "ing" words serve as participle adjectives describing one of the nouns; one "ing" word serves as a gerund.) And so we have *a sudden muffled* **sound**; we have *the* **dalliance** *of the eagles*: we have *the rushing* **contact**; we have *the clinching interlocking* **claws**; we have a *living, fierce, gyrating* **wheel**; we have **wings** and **beaks** and a *swirling* **mass**; we have **loops**; we have *a moment's* **lull** and *a motionless still* **balance**;

we have a ***parting*** (gerund); we have *their separate diverse **flight**—*but we have no verb. Every noun, though, is modified by such a vertiginous number of "ing" words, all of which give this single sentence, this single stanza, the sense of a tremendous amount of *original energy* that reaches (not unlike a strand of launched spider silk) for the memory of his experience witnessing the eagles' dalliance if not for the eagles themselves (fig. 3.4).

And still, even amid the vertiginous energy of the poem, nothing happens. A careful reader must intuit how the lack of a verb helps Whitman achieve his poetic vision. It is not merely the poem's explicit representation of "Sex and Amativeness, and even Animality" that achieves the vision; it is that the language itself epitomizes the journeywork of an energy ever reaching toward the unknown verb, not unlike how the energy of this momentary union between eagles is a fraction of a journeywork ever reaching toward the unknown of the Deep Future. The participle adjectives within the poem relate to the "buzz'd whispers" at work within a spider's abdomen. For that matter, they relate to the "buzz'd whispers" within an atom, within a cell, within an organism, within an ecosystem, within the planet, within the Milky Way, and within the stars throughout the distant "universes" of the cosmos. The single-stanza and solitary sentence of the poem further rein-

Figure 3.4. Whirling eagles, joining or releasing. Image by Robin Lowry, an Adobe Stock Contributor.

forces the idea that all ten lines are but a single note, full of intense energy, growing out of and yet contributing to the long, long arc of journeywork across the "clef of the universes."

It is possible to appreciate "The Dalliance of the Eagles" without knowing of Whitman's clef. However, if we as readers truly *hear* the poem (which is a *song*)—and if we are able to place the poem/song and the eagles within the context of Whitman's clef and the journeywork of original energy across Deep Space and Time—we have a much more humbling and staggering experience of the poem. We may join Whitman with his sense of epic astonishment at the mating eagles. Our thoughts may float to the "buzz'd whispers" of energy at work during the "moment's lull" between the eagles, and yet also to the "buzz'd whispers" at work in every cell and every atom of matter making up the eagles' bodies in the first place, and yet also to the "buzz'd whispers" within each of the participle adjectives in all their verbal energy. "The Dalliance of the Eagles" can be seen as an homage to "animality," but it is an "animality" that is a hyperobject, "massively distributed" throughout the cosmos. Whitman grants us the gift of a brush encounter with such energy by infusing his ten lines with an intense amount of the "buzz'd whispers" of language.

Why Read Whitman to Understand an Animal

We may not ever be so fortunate to witness the epic dalliance of eagles within the atmosphere. However, Whitman's work on the megafauna of the planet ought not eclipse his attentiveness to the little critters and insects (and spiders) that burrow or flit their way through his work. The prose pieces "A July Afternoon by the Pond" and "Locusts and Katydids" show Whitman directing just as much attention to the "original energy" emanating from crickets, grasshoppers, locusts, katydids, water snakes, and dragonflies "with wings of lace, circling and darting and occasionally balancing themselves quite still, their wings quivering all the time" (*Specimen Days*).

Whitman did not have access to our twenty-first-century sources concerning insects, pachyderms, whales, primates, or birds. He never heard of Alex the Parrot's conversations. He never saw the YouTube video of Robin Williams in a tickle war with Koko the Gorilla. He could not, like anyone with a phone today, search, find, and then share a dazzling image of a dragonfly with others, near and far, in a matter of seconds (fig. 3.5). During his time, photography was only emergent. Many people may never think to turn to Whitman when asking the question, *what is an animal?* After all, we can so readily turn to social media to see and share stunning images of dragonflies and other species, or we can type a few words and find scientific articles sharing recent breakthroughs in our understanding of innumerable species on this shared planet. Such images and knowledge—so readily available to us, so easily circulated through contemporary culture—can help us see and understand the dazzling architecture and the exquisite form of an invertebrate's wings and body.

However, those who *have* read Whitman are also able to place a dragonfly on the "clef of the universes." They can recognize the profundity of an encounter with the original energy at work across the cosmos even as it manifests itself through the body of a flying insect. They can learn through some quick, online research that the family of dragonflies has been around for three hundred million years living out a journeywork that has survived multiple extinction events. They can reflect upon the Emersonian idea that thought is prior to form—that the spirit of a plant or an animal is prior to the emergence of the architecture of the plant or animal. And finally, they

Figure 3.5. A close-up image of a dragonfly is an opportunity to celebrate the species' architecture of form and to hold the animal up on the "clef of the universes," marveling at the arc of *vitality*, of *animality*, of the *original energy* necessary to bring forth form. This image juxtaposes the form of the branch in its becoming with that of its "mirror," the winged-beast. Both branch and insect manifest exquisite form. Image by Virat, an Adobe Stock Contributor.

can recognize that the original and vital energy that brought forth the form of the dragonfly is also at work bringing forth the forms of language, including the long, long arcs of Whitman's life project, *Leaves of Grass*, from the title, down through every poem he drafted, included, revised, and included anew; through every section, stanza, line; through every enjambed sentence; on through to the very last clause of the 1891–92 *Leaves of Grass*: "the strongest and sweetest songs yet remain to be sung" (*LG* 1891–92).

And yet, with Whitman, there is no "final word." The fact of the matter is that Whitman's entire project becomes nothing short of a repository, not unlike the abdomen of a spider, not unlike the single stanza of "The Dalliance of the Eagles," full of the labial whispers of original energy across all atomic and yet cosmic magnitudes. *Leaves of Grass* holds whispering poems; which hold whispering stanzas, sentences, and lines; which hold phrases and words; which hold letters holding sounds—all surrounded by empty gaps and spaces holding cosmological silence. He did not have the last word, for *Leaves of Grass* is best described as *musing, venturing, throwing, seeking* as it launches forth its *filaments* of language, still reaching its strands of sound into a Deep Future only to *catch somewhere* within the neural networks of a reader's consciousness or around the curl of a reader's speaking tongue.[5] The poet's work, so closely akin to the work of a spider, is nothing short of the whispering work of Animality.

When we see an animal in the context of the "clef of the universes," we gain a perspective that yanks us out of an anthropocentric gaze. The human remains something holy, but suddenly we perceive the innumerable notes resounding in and through and across Deep Space and Time. The experience of Whitman's clef certainly has a lasting significance. Every new fact of—and encounter with—another species of this shared planet can now be placed upon a profound clef that provides orientation. Suddenly, we, like Whitman, can listen to the "buzz'd whispers" environing us, whispers with a Deep Past and yet a Deep Future, whispers that turn a brief encounter with a dragonfly, an eagle, or spider into a sublime reflection upon a part situated within the greater and greater wholes of Whitman's clef.

Notes

1. Concerning Whitman's work, one must note how he continually expanded, shifted, revised, and published his ongoing (and growing) collection of poems across multiple decades. Each publication shares the same title, *Leaves of Grass*. The Walt Whitman Archive provides online access to each publication of *Leaves of Grass*. This makes the exploration of the evolution of *Leaves of Grass* readily available and even searchable. For ease and clarity of citation, I abbreviate *Leaves of Grass* (*LG*) and then include the year of the edition's publication from which I quote.

2. My use of the term "brush encounter" comes from a merging of the ideas from Franklin Ginn's introduction, "Unexpected Encounters with Deep Time," and Peter Feng's work exploring hyperobjects and W. S. Merwin's poetics of encounter and yet withdrawal in *Desire and Infinity in W. S. Merwin's Poetry*. We may experience a fraction of the cosmological energy before it recedes out of our reach. Like bumping into a passerby on the street, the brush encounter with original energy may be brief and unexpected. In such an approach, synecdoche, the fraction of the whole, reigns supreme.

3. Whitman's focus on the male capacity to make sperm does not reduce his overall vision to androcentrism. Far from it. As will be discussed later, labial whispers and murmurs are very much at the heart of his cosmic vision.

4. Interestingly, Melville, Dickinson, Thoreau, and Whitman all integrate the atom in, respectively, *Moby-Dick; or, The Whale*, multiple poems, *Walden*, and *Leaves of Grass*. Even though they were limited by their historical circumstances, all four recognized the importance of the dynamic relationship between the part (the atom) and the whole (the cosmos).

5. The idea of the ever-reaching tendrils of Whitman's book is not mine but was shared with me by Ed Folsom during an informal conversation on a train.

Works Cited

Barlow, Roger. "If Atoms are Mostly Empty Space, Why Do Objects Look Solid?" Phys.org., February 16, 2017. https://phys.org/news/2017-02-atoms-space-solid.html.

Crawford, Allen. *Whitman Illuminated: Song of Myself*. Portland, OR: Tin House Books, 2014.

Emerson, Ralph Waldo. *The Essential Writings of Ralph Waldo Emerson*. Edited by Brooks Atkinson. New York: Modern Library, 2000.

Feng, Dong. *Desire and Infinity in W. S. Merwin's Poetry*. Baton Rouge: Louisiana State University Press, 2021.

Folsom, Ed. "Whitman Making Books/Books Making Whitman: A Catalog and Commentary." The Walt Whitman Archive. Lincoln: Center for Digital Research in the Humanities, University of Nebraska, 2005. https://whitmanarchive.org/criticism/current/anc.00150.html.

Ginn, Franklin, Michelle Bastian, David Farrier, and Jeremy Kidwell. "Introduction: Unexpected Encounters with Deep Time." *Environmental Humanities* 10, no. 1 (May 2018): 213–25.

Hoffmeyer, Jesper. *Biosemiotics: An Examination into the Signs of Life and the Life of Signs*. Scranton, PA: University of Scranton Press, 2008.

Morton, Timothy. *Hyperobjects: Philosophy and Ecology after the End of the World*. Minneapolis: University of Minnesota Press, 2013.

"Murmur." *Online Etymology Dictionary*. Edited by Douglas Harper, 2023. https://www.etymonline.com/search?q=murmur.

Panksepp, Jaak, and Lucy Biven. *The Archaeology of Mind: Neuroevolutionary Origins of Human Emotions*. New York: Norton, 2012.

Wheeler, Wendy. "The Biosemiotic Turn: Abduction, or, the Nature of Creative Reason in Nature and Culture." In *Ecocritical Theory: New European Approaches*, edited by Axel Goodbody and Kate Rigby, 270–82. Charlottesville: University of Virginia Press, 2011.

Whitman, Walt. *Leaves of Grass* and *Specimen Days*. The Walt Whitman Archive. Edited by Matt Cohen, Ed Folsom, and Kenneth M. Price. Lincoln: Center for Digital Research in the Humanities, University of Nebraska, 2022. https://whitmanarchive.org/.

CHAPTER 4

Emily Dickinson's Geographic Imagination

SUSAN L. ROBERSON

The Missing All—prevented Me
From missing minor Things.
If nothing larger than a World's
Departure from a Hinge—
Or Sun's extinction, be observed—
'Twas not so large that I
Could lift my Forehead from my work
For Curiosity (#985)

Except for a few occasions, Emily Dickinson spent her life in Amherst, a rural university town set in "a landscape of hills and streams, wildflowers and fields" (McDowell 19) in west-central Massachusetts. Most of that time she remained at the family home, the Homestead, with her parents and sister, Lavinia; her brother, Austin, and his family were close by just across the hedge at the Evergreens (fig. 4.1). The Homestead, a fourteen-acre property, was replete with gardens, orchards, fields, and woods and a conservatory her father had built so that the avid gardener could house her bedding plants and more fragile flowers. Even her bedroom was a veritable garden with its rose-patterned wallpaper. So, it is no surprise that her geographic imagination is shaped by the local environment, by the small, familiar creatures of her home grounds. The close and personal, the mundane and domestic—the "minor Things"—rather than the vast and distant make up her geographic imagination. Looking at the natural world from a domestic, woman's standpoint, she deploys a geographic "poetics of detail" (Fetterley and Pryse 259) to describe what she sees from her "window pane" (#327) or on her woodland jaunts with her dog Carlo. Focused on what she could see, Dickinson's geographic imagination generates metaphors for the local environment, for seeing "New Englandly" (#285) and thinking about the observing self (Slovic 3–4). Writing a local geography alive with motion and movement, she creates metaphors of mobility to imagine the freedom to soar intellectually and emotionally, to cross boundaries of art, being, and thought. As well, the regional geographic imagination provides a vantage point from which to interrogate power relations worked out in the natural world and the forces, "The Missing All" (#985), that propel it. While she is grounded in the local environment, her mind and imagination wander afield to oceans of thought, to subversive, contrarian, skeptical ways of accounting for the world. From her vast oeuvre, with its "multifarious imaginations, observations, and perceptions of the natural world" (Chen), I have chosen a handful of poems to illustrate how she uses animals in constructing a geographic imagination that exhibits multiple ways of thinking about the environment and the animals that inhabit it.

Dickinson had a keen eye for the natural world and the animals that resided within it, even if that world by and large existed in the domestic spaces of home and village. Her letters often mention the flowers and animals that inhabit her world, and in nearly seven hundred of her poems (about 40 percent) we find references to a variety of animals, from birds and bees to snakes and flies (Fraser). Even though the animals and flora of New England populate many of her poems and she constructed an herbarium of local flora, we would not classify Emily Dickinson as a nature writer in the strictest sense or as an environmentalist, even if there are moments when she advocates for the preservation of the natural environment and respect for natural creatures and "engages with a number of proto-ecological concerns" (Gerhardt 294). While her accounts of the natural world exhibit "a passion for metaphors, patterns, feelings, and self-awareness," her descriptions lack the objectivity of the scientific method that marks classic nature writing (Stewart xv–xvi). Nor does she write about the natural world to produce "an environmentalist commitment," as Lawrence Buell argues Henry David Thoreau and John Muir do (138). Rather, we can think of her play with the natural world as an example of a regional geographic imagination because of the ways she creates a space where metaphor and fact meet. Defining the geographic imagination, Jen Jack Gieseking notes, "While geography appears to focus primarily on the material, the imagination . . .

opens up questions of the abstract, creative, and possible. The geographical imagination affords the user ways to pry open the power in assumptions, stereotypes, and expectations associated with space and place, and to delve into how and why they are linked" (2). Coupled with careful, perceptive observation, metaphor, James Olney reminds us, is another "way of knowing" by which to "grasp the unknown and thereby fit that into an organized, patterned body of experiential knowledge" (qtd. in Slovic 75), just as we see Dickinson doing when she deploys the "minor Things" of nature to ponder the "Missing All." And because she perceives her world "New Englandly," hers is also a regional geographic imagination. Tom Lynch claims that in addition to "establishing a particular place delineation," a bioregionalist stance implies a "cultural and political practice and an ecologically based form of place-conscious self-identity" (18). Similarly, in their study of women's regionalist writing, Judith Fetterley and Marjorie Pryse define region "less as a term of geographical determination and more as discourse or a mode of analysis, a vantage point within the network of power relations that provides a location for critique and resistance" (11). When Dickinson claims that "The Robin's my Criterion for Tune" (#285) and that she sees the world and "discerns" meaning "Provincially," she evinces a bioregionalist stance. Yet, in the end, her vision is not just about New England, for as Yi-Fu Tuan has argued, the ways humans view geographic space tells us about their psychological, social, and spiritual orientations and the mythical, pragmatic, and abstract meanings they impose on the geography (58, 17). This is what Dickinson does when she contemplates the local animals and what they represent in terms of creativity, imagination, being, and power relations in social, natural, and cosmic ecologies. Beginning with her poetic character sketches of local animals, this chapter examines the ways Dickinson constructs a geographic imagination that uses these creatures to think about the environment, to inspire intellectual and creative mobility and freedom, and to interrogate cultural assumptions about religion and the forces of nature.

Birds, Bees, and Other Familiar Characters of Dickinson's Geographic Imagination

While the desert's leopard and a few other nonindigenous animals, such as the nightingale, make appearances in her poetry, local creatures account for most of Dickinson's animal references. Like the owl in "The Judge is like

Figure 4.1. Austin Dickinson house, Amherst, Massachusetts, July 2006. Photo by Daderot. Wikimedia Commons (CC BY-SA 3.0).

the Owl" (#699) who resides in the oaks near the family barn and whose rent is paid with "His favorite Refrain," Dickinson's animals are primarily familiar characters, ones she would encounter going about the grounds of the Homestead. According to Jonathan Skinner in his review of *A Spicing of Birds: Poems by Emily Dickinson*, edited by Jo Miles Schuman and Joanna Bailey Hodgman, 222 of her poems mention birds, with 26 species named; all but one mention of the nightingale are of local "backyard" birds: robin, 38; bobolink, 12; sparrow, 9; jay, 7; hummingbird, 5; crow and oriole, 4; phoebe and wren, 3; blackbird, 2 (107). Indeed, Schuman and Hodgman claim that Dickinson "falls into the categories of bird watcher and bird lover" and knew the habitats and the "unique vocalizations" of individual species (xiii, xix). In addition to multiple mentions of birds, Judith Farr claims that Dickinson wrote about the butterfly "more than fifty times," just less than the "pollinating bee," including both in her playful doxology in #18, "In the name of the Bee— /And of the Butterfly— /And of the Breeze—Amen!" (145–46). Along with the bees, birds, and butterflies, others of "Nature's People" (#986) in her neighborhood—flies, spiders, and garter snakes—attract her attention, becoming actors in some of her best-known poems.

Like a nature writer, she pays meticulous attention to the natural world in her poetry. Some of her most delightful poems are character sketches in which she brings to life the "minor" creatures, giving them the respect of her attention and situating them in a kinship community with herself that eschews an "imperial" view of nature (Worster 83–85, 29). Her sketch of a house spider in "The Spider holds a Silver Ball" (#605), most likely describing the common American house spider who unwinds "His Yarn of Pearl" and "Supplants our Tapestries with His," exemplifies both the interrelatedness of the natural and human worlds and the "poetics of detail" that characterize women's regionalist writing. In this poem she deftly describes how the spider by means of "unperceived Hands" plies his "Trade" of making webs, tapestries that vie with those made by human hands. Although the cobwebs of the American house spider resemble a "tangled mess" ("4 Most Common New England Spiders"), Dickinson imagines them as tapestries, carefully woven hangings that tell a story. Before screened windows were the norm, spiders and flies could easily invade the domestic space, as the spider's "Continents of Light / . . . dangle from the Housewife's Broom— / His Boundaries—forgot—." Paying attention to the small animals, she demonstrates how "the infinite particularity of women's lived experience"—like the cobwebs that the broom works to sweep away—become "'practical images' that serve as 'symbolic vehicles'" and a "vantage point" from which to assess relations of power and knowledge (Fetterley and Pryse 259, 11). In this poem where geographic description and imagination meet, she creates a space where her account of the spider prompts her to think about the value of work—the spider's work of weaving and woman's work of cleaning house. She also uses the spider's web as a symbolic vehicle by which to think about art—the spider's creation, women's woven artforms, and poetry, another gossamer form that can forget "Boundaries" and disrupt quotidian assumptions.

As in the poem about the artful spider, Dickinson demonstrates the method of the geographic imagination as she brings other regional creatures to life and invests them with meaning at the human, environmental, and cosmic levels. Poem #1405 describes bumblebees, "Black, with Gilt Surcingles" as "Buccaneers of Buzz" whose honey "Jugs" made from nectar gathered from the local flowers are part of the "Universe's" ecology. Figured as pirates wearing black belts who subsist on the "Fuzz" of flowers, which her lexicon likens to pollen and plant nectar, the bees in this poem evince Dickinson's understanding of the ways ecosystems depend on multiple players to sustain life ("Lexicon"). Although they do not produce the kind of honey we might enjoy with our morning toast (that is the work of honeybees), bumblebees do create waxen "honey pots" to store the nectar and pollen that will sustain the queen bee, her larvae, and the new worker bees for their annual life cycle. In turn, as the avid gardener understood, they help pollinate the plants they visit as they gather their nectar ("Do Bumble Bees").

In #1483 she compares the robin to the "New England Farmer," writing that "His Dress denotes him socially, / Of Transport's Working Classes." With their "portly persons and warm red jackets," as Susan Fenimore Cooper described them (11), American robins, like farmers, are stalwarts of the meadows and fields of rural New England, live in "humble circumstances," and are characterized by "punctuality" and "integrity." Yet the robin and the farmer are both figured as fugitives "Cajoling Consternation By Ditties to the Enemy," as if they had to sing away their fears. In what is otherwise a description of commonplace robins and humble farmers, the poem punctuates their commonality when Dickinson situates them in natural and cosmic ecologies that require strategies to deal with life's harshness.

In another poem that shows the play between careful observation and metaphor, she compares the jay to a "Brigadier," its wing and crest feathers blue like the uniform of an American soldier (fig. 4.2). Capturing the "chaffing, scolding, scoffing" call of the jay (Ackerman 107), she describes how with his song "Brittle and Brief," he sits upon a bough, "Confident and straight," marshaling in "March" and the coming spring (#1177). Playing with military imagery, she suggests that March must march through the forces of winter to make way for spring as the jay oversees the changing of the seasons.

Even the caterpillar gets attention when she describes "How soft a Caterpillar steps" and the "velvet world" from which it comes (#1448). Examining the caterpillar in her hand, she remarks how "Its soundless travels just arrest / My slow—terrestrial eye" that thinks primarily of "its own career / What use has it for me—." This moment demonstrates her attentiveness to detail—the "soundless travels"—and ways of seeing nature, of looking

Figure 4.2. *Crested Blue Jay (Pica cristata caerulea)*, by Mark Catesby. Mark Catesby and George Edwards, *The Natural History of Carolina, Florida and Bahamas*, vol. 1, pl. 15, 1754. Wikimedia Commons.

and listening patiently and thinking about the correspondences between the animal and human worlds. Perceiving the relation between the caterpillar and herself, she quickly and quietly disarms the use-value argument about nature. In an example of environmentalist thinking, she questions assumptions that nature is supposed to serve the needs of humans, even if only symbolically as figures of metamorphosis and immortality.

In probably her most famous bird sketch, she describes the rush and color of a ruby-throated hummingbird (#1463). In quick strokes she paints "A Resonance of Emerald— / A Rush of Cochineal" that visits "every Blossom on the Bush." In constant action as it rushes from plant to plant, beating its wings "up to seventy-five times a second" (Ackerman 214), this migratory hummingbird has traveled from afar as it brings "The mail from Tunis, probably." Even though the hummingbird typically migrates to Central America, Mexico, and Florida for the winter, Dickinson imagines it journeying from the north coast of Africa (fig. 4.3). Speculating that the bird has flown from the ancient Tunisian city, the poet leaves open what the "mail" it carries might be. Perhaps it is a message of origins as the reference to Tunis would suggest, or it may refer to the natural ecological processes of migration, pollination, and continuing life. In either case, the bird carries messages of import. In nature sketches like those I have quickly looked at, Dickinson demonstrates an attention to detail, a local sense of place, and the community between herself and nature. As "A Route of Evanescence" also demonstrates, her geographic vision, though it may be local and arise from a domestic standpoint, is neither static nor narrowly focused.

The Geographic Imagination and the Freedom to Move and Think

Like the active hummingbird, Dickinson's imaginary geography is alive with motion, movement, and change. In her poems bees "Ride abroad in ostentation" sipping nectar (#1405); birds soar and shift and whirl (#653); flies interpose with "uncertain stumbling Buzz" between light and vision (#465); "Butterflies from St Domingo / Cruis[e] round the purple line" (#137); and her snake rides through the grass before it "closes at your feet / And opens further on" (#986). Seasons march through the poems, as winter like "Brooms of Steel" sweeps the "Winter Street" (#1252), and day gives way

Figure 4.3. *Ruby-throated Hummingbird (Archilochus colubris)*. Illustration by Louis Agassiz Fuertes from *The Burgess Bird Book for Children*, 1919. Wikimedia Commons.

to night as the "Sunrise runs," "The Noon unwinds," and Midnight clasps "Hemispheres" (#710). The wind coming "like a Bugle" and lightning, "The Doom's electric Moccasin," charge the landscape with beauty and danger, foretelling "How much can come / And much can go, / And yet abide the World!" (#1593). Not only do the active verbs of her nature sketches connote an environment in motion, but they indicate something about her own mobilities. Against the domestic, homebound position of her speakers, the physical mobility, movement, and motion of the natural world provide a vocabulary for translating the freedom to soar intellectually and emotionally, to cross boundaries of art, being, and thought. Travel and mobility have long been associated with new knowledge, the freedom to move, and the innovative mind "that explores new ways of looking at things," which include the material, physical world and the interior landscapes of the self (Leed 72, 13–14). Because of these qualities, travel has been associated with the subversive mind that questions the status quo and that threatens "settled patriarchal order" when women are doing the traveling (Massey 11). Although Dickinson's speakers rarely wander about, the movement of her subjects, the riding bees and soaring birds, double as metaphors for her imaginative and intellectual mobility. As she reminds us in "There is no Frigate like a Book" (#1263), the active mind can traverse realms of thought and imagination without leaving home.

Two of her bird poems evoke this freedom to move and to think, to be curious about the world and to create "Music." In "Within my Garden, rides a Bird" (#500), she follows an ever-moving hummingbird riding "Upon a single Wheel . . . Above the Ripest Rose" until he "Reels [away] in remoter atmospheres." On the surface, the poem relates the natural processes of feeding and pollination as the hummingbird partakes of the rose and then moves on. But when the poem moves from the material garden to the creative "Garden in the Brain," Dickinson links the bird's freedom to wander to "remoter atmospheres" to her own poetic curiosity. Although the speaker's dog companion, perhaps Dickinson's own Carlo, "the best Logician," looks at her askance, her eye moves "To just vibrating Blossoms!" of "Exquisite" creativity, suggesting that imagination is itself alive, vibrating, an act of movement. As nature writer Annie Dillard has observed, "The interior life is in constant vertical motion. . . . It dreams down below; it notices up above; and it notices itself, too, and its own alertness" (qtd. in Slovic 62). Likewise, Dickinson's poem about curiosity, creativity, and thought follows not the dull logic of the mundane world but the "dizzy Music" of the bird in flight. Then, in poem #653, she identifies "Being" as a "Bird." Just as the bird "soars—and shifts—and whirls— / And measures with the Clouds," so the self, Being, soars afloat "The General Heavens" at an "Easy—even—dazzling pace." Free to soar the heights of air and thought, Being experiences music, poetry, just "For Ecstasy—of it," just for the pleasure and freedom of the song. Like the undulating flight pattern of small songbirds or the static soaring of the hawk, Being alternates between short bursts of flight as it rises, soars, and glides through the currents of the air (Sibley; "10 Bird Flight Patterns"). In both poems, the geographic imagination drawn from the birds of New England maps a place, a kind of paradise of the mind, and charts patterns of movement where she can think creatively and where "it is possible to say things that do not conform to the dominant culture" (Fetterley and Pryse 144). Following the actions of the birds, she imagines an innovative poet with subversive modes of thinking who soars above and beyond the everyday world and its expectations for women and poetry.

A different set of poems illustrates the links between seeing and nature, the roles of vision and visibility, of watching and awareness in Dickinson's geographic imagination. In *Writing Nature: Henry Thoreau's Journal*, Sharon Cameron has noted, "To write about nature is to write about how the mind sees nature, and sometimes about how the mind sees itself" (44). Seeing is not just an act of patiently looking at an object but also of perceiving meanings, the "Internal difference, / Where the Meanings, are" within the self (#258). While Dickinson's depictions of the small animals of Amherst testify to her careful observation, when she suffered an eye malady, she was confronted with the possibility of not seeing and what that could mean to her. Pairing #327 "Before I got my eye put out" about losing her vision with #328 "A Bird came down the Walk" helps us see how writing about nature for Dickinson is also writing about perception and being. Cynthia Griffin Wolff claims that the eye ailment she suffered between 1861 and 1865 was terrifying because she could not read; as Dickinson put it in a letter to Joseph Lyman, "It was a shutting of all the dearest ones of time, the strongest friends of the soul—BOOKS" (qtd. in Wolff 165). But "Before I got my eye put out" also suggests that looking at the natural world was important to her. Indeed, Richard Sewall asserts that gazing upon nature is an "overwhelming, even mortal, experience" akin to seeing God, enlightening and frightening as that may be (559). The poet begins by musing that before her eye ailment she "liked as well to see— / As other Creatures, that have Eyes," that is, to physically see the world around her. But after she could not see, she had to conjure up images in her mind's eye, "Between my finite eyes." When she does so, she imagines views of the natural world— the sky, the meadows, mountains, forests, "The Morning's Amber Road." Although these may be generic sites, they also resonate with those particular to Amherst, like her family's meadows or the nearby Mount Holyoke Range (fig. 4.4). Replacing physical sight with insight, however, is a dangerous proposition since "The News would strike me dead." By claiming the natural scenes and the news they convey as "mine—to look at when I liked," she asserts possession of and control over them, as an audacious poet would do. At the end of the poem the speaker opts for the "safer" choice to view the landscape from "the Window pane," perhaps from her bedroom window, which overlooked Amherst's Main Street. Yet it is actually her soul that she situates on the windowpane, making this move even more precarious, or as the poem puts it, "Incautious," to put her soul or mind in a liminal place where she might receive "The News" that nature brings.

This news, about the normal casualties of nature, is brought to her by what she sees from her window. Whether Dickinson wrote #328 on the heels of #327 or meant it to be a response to the poem, they make a good pairing here. After contemplating what she would lose if she could no longer see, the poet carefully describes "A Bird [that] came down the Walk," noting that its eyes "looked like frightened Beads" and that "He stirred his Velvet Head." She sets him in motion as he comes down the walk, bites an "Angleworm in halves," "drank a Dew," "hopped sidewise," and then nonchalantly "unrolled his feathers / And rowed him softer home." These two poems connect seeing, imagination, and perception with the natural world, one at once so far-flung that it encompasses mountains, forests, and stars and yet so close by that the actions of a bird on the walk deserve attention. The domestic vantage points of the windowpane and the walkway open imaginatively to oceans where butterflies "Leap, plashless as they swim" and where the poet can leap and swim in oceans of thought and creativity. Just as the bird's actions endanger the angleworm, often used as fish bait, so the poet's angle of vision endangers as it subverts the quotidian social world. Like the bird in a naturalistic environment who casually bites the angleworm and flies home as if nothing were untoward, the poet bites into cultural assumptions as if they were bait for thought. As these poems and the birds that figure in them

Figure 4.4. Listing of sights in Amherst, 1886. Library of Congress, Geography and Map Division, Washington, DC (Troy, NY: L. R. Burleigh, 1886), 20540-4650 USA dcu.

suggest, looking and imagining are actions, motions of the seeing and creating self. They also demonstrate that the natural world, including its animals, even from a domestic perspective, can be a location from which the poet can conjure and articulate risky, unsayable things.

Using the Geographic Imagination to Question Christian Symbology and Doctrine

Some of these unsayable things hinge on Christian symbols and doctrines of latter-day Puritanism and are often figured in the poetry by birds who bear messages from God. As poet J. D. McClatchy puts it, since "the very dawn of civilization, birds were symbols of the spirit" (qtd. in Heitman 45). Rather than deploy birds more usually associated with spiritual messages such as the dove, the owl, or the eagle, Dickinson uses the common bobolink in "Some keep the Sabbath going to Church" and an unnamed bird in "The Skies can't keep their secret!" True to the geography of New England, Dickinson deploys the realities of animal life and seasonal change in her neighborhood to charge her imagination. In her well-known pantheistic poem about finding God in nature, the speaker declares that rather than go to church on the Sabbath, she worships God in the orchard where the bobolink, the "Chorister," sings God's heavenly message (#324). The bobolink is a good choice because he looks as if he were wearing a tuxedo or a choir robe backward. The popular early nineteenth-century poet of America's natural world, William Cullen Bryant, similarly captured the bobolink's formal coat and merry song in this verse from "Robert of Lincoln":

> *Wearing a bright black wedding-coat;*
> *White are his shoulders and white his crest.*
> *Hear him call in his merry note:*
> *Bob-o'-link, bob-o'-link,*
> *Spink, spank, spink. (Bryant 486)*

In her poem about the Sabbath in which the migratory songbird warbles its message, Dickinson suggests that one need not go to church or rely on organized religion to be in the presence of God (fig. 4.5). Unlike the naturalistic poem that unflinchingly depicts the grim economy of survival in which the bird bites the angleworm in half, this charming poem echoes radical transcendentalist notions that God is present in nature. This is a theme that Ralph Waldo Emerson elaborated in his controversial 1838 "Divinity School Address" that encouraged aspiring Unitarian ministers to practice spiritual self-reliance. Opening his address by calling attention to the "refulgent summer" in which "The grass grows, the buds burst" and "The air is full of birds," Emerson looks to "the Moral Nature, that Law of laws" as a guide to spirituality rather than to the dry tenets of church doctrine (231, 237). Similarly, along with its heretical position regarding organized religion, Dickinson's poem holds out the promise that there are moral lessons to be learned from nature and that human perfection through self-cultivation, a precept of Unitarianism, is the goal of faith. As she says, "So instead of going to Heaven, at last— / I'm going, all along."

In contrast, the unnamed bird in "The Skies can't keep their secret!" (#191) does not give up its messages. Revealing her sense that the creatures of nature interact with each other, Dickinson sketches a communication chain by which the skies tell "their secret" to the hills, who tell the orchards, who tell the daffodils, one of the first flowers to reappear after winter. When a passing bird overhears the message that change is in the air, the speaker wonders if she should bribe the bird to tell it to her. At first glance it seems

Figure 4.5. *Bobolink* (*Hortulanus caroliniensis*), by Mark Catesby. Mark Catesby and George Edwards, *The Natural History of Carolina, Florida and Bahamas*, vol. 1, pl. 13, 1754. Wikimedia Commons

that the flowers and bird are announcing the coming of spring, a welcome change after a frigid Massachusetts winter. But it is telling that the bird, who would normally sing the spring in, knows a secret that it will not disclose and that the speaker claims "It's finer—not to know," perhaps because it would undermine important assumptions. When the speaker asks, "If Summer were *an Axiom*— / What sorcery had *Snow?*," she questions the logic of change, of the inevitable winter. If summer were the rule, the axiom, the given, as it might have been in Eden, then winter can only be sorcery, black magic, and the "Father" some sort of trickster who has changed the order of things "In your new-fashioned world!" Claiming not to want to know the "secret," the poet nonetheless ponders why winter (and death) intruded on the old Edenic world, subversively figuring God not as a benevolent father but as a sorcerer practicing black magic with the world. In both poems, then, Dickinson deploys animals to question assumptions about religion and the rule of God, the power dynamics worked out in the moral and natural worlds. Or, more terrifying, the possibility of the missing God, "The Missing All," as she does in her famous snake poem.

Having described the actions of "A narrow Fellow," one of "Nature's People," as it "rides" through "the Grass," Dickinson ends the sketch by commenting on the "tighter breathing" the speaker experiences upon encountering the snake (#986). Published as "The Snake" in the Springfield *Daily Republican* on February 14, 1866, the poem demonstrates her understanding of the movements and habitat of the snake, that it "likes a Boggy Acre" that is "too cool for Corn." Most likely, Dickinson is referring to the eastern garter snake, one of the most common snakes in New England (fig. 4.6). Often seen in gardens, and hence incorrectly called a garden snake, it also counts bogs and wetlands as its habitat. When she refers to the "Whip lash / Unbraiding in the Sun," she may be alluding to the long stripes that run down its body unfolding as it stretches out when the speaker stoops to catch or "secure" it. Although this snake is not venomous and is familiar, something about it triggers an emotional response, causing the speaker to feel "a tighter breathing / And Zero at the Bone." In a letter to Thomas Wentworth Higginson, Dickinson recalls that when she was a girl, she "was told that the Snake would bite me . . . but I went along and met no one but Angels" (*Letters* 178). Like the boy of the poem, the young girl did not actually fear the snake itself. Rather, there is something else that causes the speaker's reaction. Wolff contends that instead of making associations to Satan and evil or even to phallic fears, the snake's "unspeakable" terror is "reinforced by a reduction to 'Zero' within the verse" (490); it is the absence of meaning, "The Missing All," rather than physical danger that chills the speaker to the bone. Unlike those of "Nature's People" with whom the speaker has feelings of cordiality, the snake evokes a sense of fear that is not so much of the snake itself but of the lesson of nothingness, of absence, of zero, that it metaphorically portends.

Similarly, in "I heard a Fly buzz" (#465), a seemingly benign housefly that comes between the dying speaker and a view of heaven signifies zero, a lack of meaning, an undermining of a belief in an afterlife, the obliteration of the self by death that flies in the face of Christian ideas of an afterlife (Wolff 224–27). Clarence Gohdes speculates that the fly in this poem

Figure 4.6. *Garter Snake.* Drawn by Edward Knobel. *The Turtles, Snakes, Frogs and Other Reptiles and Amphibians of New England and the North,* 1896. Wikimedia Commons.

is the bluebottle fly. If that is the case, it is appropriate to the poem's theme. Bluebottle flies are distinguished by their "distinctive coloration and loud buzzing flight" ("Bluebottle Fly"), something that the speaker hears in its "uncertain stumbling Buzz." Because bluebottles commonly are drawn to carrion, the poem's fly may have been attracted to the dying body of the speaker, perhaps coming to feed on it when the speaker finally passes on. When the speaker comments, "And then the Windows failed," she may be referring to her eyes, the windows to the soul, which fail to see not just the room but the promised afterlife because of the interposing fly. She may also suggest that the window, opened in warm weather, failed to keep the fly out. As mentioned earlier, before the Civil War windows did not typically have screens to keep out pests like the bluebottle fly, which can spread disease. In any case, Dickinson uses this humble fly, something with which she would have been quite familiar, to map a geography of death and oblivion within the domestic spaces of the home. As in other poems, her objective is not so much to focus on the animal species, though her deft strokes show her familiarity with them, but to use them as part of a geographic imagination that affords her "ways to pry open the power in assumptions, stereotypes, and expectations associated with space and place, and to delve into how and why they are linked" (Gieseking 2). Here, the expectations center on the moment of death, which in nineteenth-century sentimental literature was often figured as a moment of ecstasy when heaven is revealed to the dying person, as it was to Little Eva in *Uncle Tom's Cabin*. As this pair of poems about a snake and a fly suggest, nature has its own lessons that do not cohere to those devised by organized religion or proposed by Emersonian idealism. These lessons follow a more ruthless, Darwinian view of nature that humans "had not been created with special care in the image of God" but rather are "one with all other species in a universal brotherhood of living and dying" (Worster 180).

Another assumption that she interrogates is the promise of spring and the resurrection commonly associated with it. This is something Edward Hitchcock, the renowned president of Amherst College, geologist, theologian, and friend of the Dickinson family, preached in *Religious Lectures on Peculiar Phenomena in the Four Seasons* (1850). Richard Sewall summarizes, "Throughout the *Religious Lectures*, and in a hundred different ways, Hitchcock stressed the Christian theme of the Resurrection. Life is imperishable, its seed never extinct. Winter is but a harbinger of spring" (348). The colored frontispiece of the *Lectures* engraved by his wife, Orra White Hitchcock, an early female scientific illustrator, features "an idealized landscape, a little pond with trees and bushes, all burgeoning with the emerging life of spring—a butterfly coming from a cocoon, a frog from pollywogs in the pond, a bird hovering about her nest, and spring flowers all about. The caption under the picture reads, "Emblems of the Resurrection" (Sewall 348). The scene Orra Hitchcock paints is full of life emerging from the stasis of winter (fig. 4.7).

Similarly, Dickinson often portrays the coming of spring with images of a natural world alive with activity and transformation after the long winter, as she does in the early poem "Some Rainbow—coming from the Fair!" (#64). Against a sky colored by a rainbow and peacock purples, the spring animals "resume the whir" of "last year's sundered tune!" Punctuated by verbs, the poem imagines the returning life: "The dreamy Butterflies bestir," the "Baronial Bees—march—one by one—," "The Robins stand as thick today / As flakes of snow stood yesterday," "The Orchis binds her feather on," and "Regiments," perhaps of "Turkish" tulips, stand at attention in the warming weather (Farr 132–33). In a letter of April 1856 to John Graves, she wrote of the emerging spring in similar terms: "And here are Robins— just got home—and giddy Crows—and Jays—and will you trust me—as I live, here's a *bumblebee*" (*Letters* 135). Capturing the ways that bumblebees

Figure 4.7. *Emblems of the Resurrection*, by Orra White Hitchcock. Edward Hitchcock, *Religious Lectures on Peculiar Phenomenon in the Four Seasons*, Edward Hitchcock, 1850. Used with the permission of Archives and Special Collections, Amherst College.

fly in groups and spring robins emerge from solitary positions of wintering in the trees to populate the meadows and yard, the poem evinces her understanding of the natural world's emergence from winter. Yet, even as this poem portrays the resurgence of life, its militaristic language, Wolff contends, drains the poem of "transcendent implication" (434), asking one to question whether it can be an emblem of the resurrection if the "Regiments of Wood and Hill" stand "Without Commander!," without a God who orchestrates the natural world. Similarly, in the letter, the "gay" activity of the spring reminds her of "*sadder* features" of death and disappearance, "a mouldering plume, an empty house, in which a bird resided" (*Letters* 135), the inexorable cycles of life and death that are the "product of blind physical laws" (Worster 122) rather than the designs of an overseeing God. In these early contemplations about spring and what it portends, we catch glimpses of the ways Dickinson uses her geographic imagination to question assumptions about accepted Christian symbology and doctrine. Although she consistently writes of the coming of spring, the birds who "reported from the South," the blooming flowers and visiting bees, nonetheless, the "frosts" (#743) of winter and death reappear, always "the punctual snow!" (#99), belying the promise of resurrection or an Eden. As she writes to Graves, "It is a jolly thought to think that we can be Eternal—when air and earth are *full* of lives that are gone—and done—and a conceited thing indeed, this promised Resurrection!" (*Letters* 135).

"A Pang is more conspicuous in Spring" (#1530) likewise suggests that in contrast to the "Minute Effulgencies," the splendors of spring and "the things that sing," there is ironically a "Pang" that is more conspicuous at this season. Evoking the songs of "Birds" and "a Blue Bird's Tune," she reminds readers of the early birds of spring who sing it in. The eastern bluebird, which flourished during the eighteenth and nineteenth centuries as settlers cleared the forests, was associated with the coming of spring. Because of this connection with spring and the sprightliness of their song, bluebirds represent joy and hope ("Eastern Bluebird"). By clarifying, however, that it is "Not Birds entirely—but Minds" that sing, the poem suggests that like the bluebird, the poet joyfully sings in the spring. Yet by connecting the "things that sing" with "A Pang" instead of joy when she claims that "Resurrection had to wait / Till they moved the Stone," she questions the promises of spring and the resurrection it represents. Painfully she asks, "Who cares about a Blue Bird's tune" when "what they sung for is undone"; perhaps by the winter that will surely come again. Additionally, we are asked to infer that "a Blue Bird's Tune" is broken because the resurrection did not happen of itself or by the hand of God; it had to be manufactured by someone, the "they" in the last line of the poem. When she ends, "Why, Resurrection had to wait, / Till they had moved a Stone—", she questions the bedrock miracle of Christianity, implying that theologians invented the story of the resurrection and hence of Christianity. In these poems, birds who are figured as harbingers of spring and the possibilities of resurrection also remind us of the winter that will return, of the "Missing All" if the story of the resurrection is just a story, if there is no "Commander" but only the laws of nature marshaling in the seasons. As Nina Baym argues about Dickinson's use of science, her "overriding poetic projects . . . are to query the existence of 'heaven' . . . and to dismantle" claims about a beneficent God (147).

The Robin's Tune and Dickinson's Bioregionalism

Winter also figures in "The Robin's my Criterion for Tune" (#285). Countering the "lie" of English literature about winter, she conceives of it as "Snow's Tableau," a snow-covered landscape with its own drama that informs her

86 | CHAPTER 4

Figure 4.8. *Robin*, by John James Audubon. John James Audubon Letters and Drawings, MS Am 21 (46), Houghton Library, Harvard University.

vision and the questions that she raises. Comparing the geographies of New England and Old England, she announces that whether one is the queen of England or herself, she "discerns . . . Provincially." She proclaims that she prefers the robin to the cuckoo, and the buttercup and daisy to England's "Whim for Bloom," perhaps the Tudor Rose, the national flower of England (fig. 4.8). Writing that she is "Orchard sprung," she spins a more homely image reminiscent of the fruits of the Dickinson orchard, its apples, pears, and quinces. When she mentions that "the Nut—October fit" signals the change of season when it drops from the tree, she may be thinking about local acorns, black walnuts, and chestnuts (McDowell 141–47). As she consistently does in the poems about animals and the natural world, Dickinson situates herself in the New England bioregion and indicates how it has shaped her imagination, perception, and identity.

Moreover, when she announces at the beginning of the poem that she sees the way she does "Because I grow—where Robins do—," she indicates the role that the common robin has in shaping her imagination and poetry, her "tune." Mentioned in her poetic oeuvre thirty-eight times, more than any other bird, the robin relays what it means to see "New Englandly" (Skinner 107). The American robin, distinct from the English robin, is the quintessential American bird. It is the most abundant bird in North America, associated with the coming of day and spring and with industry and luck, features that make it an apt symbol of the New Englander. Claiming the robin as the symbol of her New England upbringing, she also posits it as her criterion for tune, her principle for birdsong and for poetry. Jennifer Ackerman tells us that "songbirds go through the same process of vocal learning that people do—they listen to adult exemplars, they experiment, and they practice, honing their skills like children learning a musical instrument" (141). Though most known for its cheery song welcoming the day, the robin has a medley of songs to communicate varied messages to careful listeners, much as the poet does. As Ackerman explains, birds vocalize for a variety of reasons, "to warn off predators and to identify family, friends, and foes" as well as "to defend their territory . . . and to woo a mate" (141). Similarly, Emily Dickinson learned her craft and honed her skill as a poet to communicate a range of perceptions and messages. Ackerman further explains that bird calls "are typically short, simple, succinct, and innate," while songs are generally "longer, more complex, and learned" (141). Ackerman's description of bird songs and calls corresponds uncannily to Dickinson's poetic method—her short, staccato lines and complex, learned allusions. By claiming that "The Robin's my Criterion for a Tune," Dickinson is implying that not only is she molded by her geographic region but also that her verse may be informed by the bird songs and calls to which she carefully attends. Like the robin, a herald of new days and new seasons as well as new ways of thinking, she also sings out alarms, exposing the "lies" that "flit" over the truths of nature and life, of the "Missing All," and the power relations worked out in the natural world. After describing bird calls and songs, Ackerman claims, "Singing is both risky and expensive" because singing birds are exposed to predators and sacrifice time from foraging to make music (161). We could also say that writing poetry, for Dickinson, like singing, is "both risky and expensive," emotionally and intellectually, which may be one reason that this innovative American poet did not publish more poems during her lifetime.

Few contemporaries understood or appreciated her experimental form and sometimes troubling messages. In response to Thomas Wentworth Higginson's "Letter to a Young Contributor" in which he advises young writers in the *Atlantic Monthly* (April 1862), Dickinson asked of the noted writer and advocate of social causes, "Are you too deeply occupied to say if my

Verse is alive?" (*Letters* 171). Yet she seems disappointed in his response to her poems, in which he judges her style of singing, its rhythm and complex, subversive imagination, as too "risky" to publish. Understanding that her poetry flies in other realms as if she were a fish out of water, she rejoins:

> I smile when you suggest that I delay "to publish"—that being foreign to my thought, as Firmament to Fin—
> . . .
> You think my gait "spasmodic"—I am in danger—Sir—
> You think me "uncontrolled"—I have no Tribunal. (*Letters* 174)

Deploying animal metaphors even here in the letter to Higginson, she illustrates the geographic imagination and the creative and intellectual risks that make her one of our classic poets, one that uses perceptive sketches of local, New England animals to create a space where she has the mobility and freedom to interrogate the meanings of poetry, the self, and religion. Joining metaphor and fact, Dickinson shapes a geographic imagination populated by the familiar creatures of her home grounds as she investigates such universal themes as creativity, being, and the "Missing All."

Note

All poems refer to the numbers assigned in *The Complete Poems of Emily Dickinson*, edited by Thomas. H. Johnson (Little, Brown, 1960). Available in paperback, this standard edition of Dickinson's poetry is readily accessible to readers. A more recent edition of the poems, The Variorum Edition of *The Poems of Emily Dickinson*, edited by R. W. Franklin (Belknap Press of Harvard University Press, 1998), is an excellent resource for researchers. Readers can also access the Franklin edition and manuscript copies of the poems at the Emily Dickinson Archive, https://www.edickinson.org/.

Works Cited

"10 Bird Flight Patterns to Know." Bird Protection Quebec. Accessed November 15, 2022. https://pqspb.org/bpqpoq/10-bird-flight-patterns-to-know/.

"4 Most Common New England Spiders." Catseye, February 22, 2022. https://www.catseyepest.com/pest-facts-and-prevention-101/common-new-england-spiders/.

Ackerman, Jennifer. *The Genius of Birds*. New York: Penguin, 2016.

Baym, Nina. *American Women of Letters and the Nineteenth-Century Sciences: Styles of Affiliation.* New Brunswick, NJ: Rutgers University Press, 2002.

"Bluebottle Fly." *Britannica.* Accessed November 15, 2022. https://www.britannica.com/animal/bluebottle-fly.

Bryant, William Cullen. "Robert of Lincoln." In *The American Tradition in Literature*, vol. 1, 3rd ed., edited by Sculley Bradley, Richard Croom Beatty, and E. Hudson Long, 486–88. New York: W. W. Norton, 1967.

Buell, Lawrence. *The Environmental Imagination: Thoreau, Nature Writing, and the Formation of American Culture.* Cambridge, MA: Belknap Press of Harvard University Press, 1995.

Cameron, Sharon. *Writing Nature: Henry Thoreau's Journal.* Oxford: Oxford University Press, 1985.

Chen, Grace Mei-shu. "Coda: Natural Messages and Aesthetic Pleasure in Emily Dickinson's Nature Writing." In *Emily Dickinson's Lyrical Ecologies: Forays into the Field*, edited by Marta Werner and Eliza Richards. Dickinson Electronic Archives, 2013. http://www.emilydickinson.org/emily-dickinson-lyrical-ecologies-forays-into-the-field.

Cooper, Susan Fenimore. *Rural Hours.* Edited by Rochelle Johnson and Daniel Patterson. 1850. Reprint, Athens: University of Georgia Press, 1998.

Dickinson, Emily. *The Complete Poems of Emily Dickinson.* Edited by Thomas H. Johnson. Boston: Little, Brown, 1960.

———. *Selected Letters.* Edited by Thomas H. Johnson. Cambridge, MA: Belknap Press of Harvard University Press, 1958.

"Do Bumble Bees Make Honey?" BuzzAboutBees.Net, The Wonderful World of Bees. Updated November 2023. https://www.buzzaboutbees.net/Do-Bumblebees-Make-Honey.html.

"Eastern Bluebird." Wikipedia. Accessed November 15, 2022. wikipedia.org/wiki/Eastern_bluebird.

Emerson, Ralph Waldo. "Divinity School Address." In *Transcendentalism: A Reader*, edited by Joel Myerson, 230–46. Oxford: Oxford University Press, 2000.

Farr, Judith, with Louise Carter. *The Gardens of Emily Dickinson.* Cambridge, MA: Harvard University Press, 2004.

Fetterley, Judith, and Marjorie Pryse. *Writing out of Place: Regionalism, Women, and American Literary Culture.* Urbana: University of Illinois Press, 2003.

Fraser, Alison. "'Buccaneers of Buzz': Dickinson's Humanimal Poetics." In *Emily Dickinson's Lyrical Ecologies: Forays into the Field*, edited by Marta Werner and Eliza

Richards. Dickinson Electronic Archives, 2013. http://www.emilydickinson.org/emily-dickinson-lyrical-ecologies-forays-into-the-field.

Gerhardt, Christine. "'Often seen—but seldom felt': Emily Dickinson's Reluctant Ecology of Place." In *Critical Insights: Emily Dickinson*, edited by J. Brooks Bouson, 292–321. Hackensack, NJ: Salem Press, 2010.

Ghodes, Clarence. "Emily Dickinson's Blue Fly." *New England Quarterly* 51, no. 3 (1978): 423–31.

Gieseking, Jen Jack. "Geographical Imagination." In *International Encyclopedia of Geography: People, the Earth, Environment, and Technology*, edited by D. Richardson, N. Castree, M. Goodchild, A. Jaffrey, W. Liu, A. Kobasyashi, and R. Marston, 1–8. New York: Wiley-Blackwell and the Association of American Geographers, 2017.

Heitman, Danny. *A Summer of Birds: John James Audubon at Oakley House*. Baton Rouge: Louisiana State University Press, 2008.

Leed, Eric J. *The Mind of the Traveler: From Gilgamesh to Global Tourism*. New York: Basic Books, 1991.

"Lexicon." Emily Dickinson Archive. Accessed January 9, 2023. https://www.edickinson.org/words.

Lynch, Tom. *Xerophilia: Ecocritical Explorations in Southwestern Literature*. Lubbock: Texas Tech University Press, 2008.

Massey, Doreen. *Space, Place, and Gender*. Minneapolis: University of Minnesota Press, 1994.

McDowell, Marta. *Emily Dickinson's Gardening Life: The Plants and Places That Inspired the Iconic Poet*. Portland, OR: Timber Press, 2019.

Schuman, Jo Miles, and Joanna Bailey Hodgman, eds. *A Spicing of Birds: Poems by Emily Dickinson*. Middletown, CT: Wesleyan University Press, 2010.

Sewall, Richard B. *The Life of Emily Dickinson*. New One-Volume Edition. New York: Farrar, Straus and Giroux, 1974.

Sibley, David. "Identifying Small Song Birds by Flight Styles." Sibley Guides, March 11, 2011. https://www.sibleyguides.com/2011/03/identifying-small-songbirds-by-flight-style.

Skinner, Jonathan. "Birds in Dickinson's Words." Review of *A Spicing of Birds: Poems by Emily Dickinson*, edited by Jo Miles Schuman and Joanna Bailey Hodgman. *Emily Dickinson Journal* 20, no. 2 (2011): 106–10.

Slovic, Scott. *Seeking Awareness in American Nature Writing: Henry Thoreau, Annie Dillard, Edward Abbey, Wendell Berry, Barry Lopez*. Salt Lake City: University of Utah Press, 1992.

Stewart, Frank. *A Natural History of Nature Writing*. Washington, DC: Island Books/Shearwater Books, 1995.

Tuan, Yi-Fu. *Space and Place: The Perspective of Experience*. Minneapolis: University of Minnesota Press, 1977.

Wolff, Cynthia Griffin. *Emily Dickinson*. Boston: Addison-Wesley, 1988.

Worster, Donald. *Nature's Economy: A History of Ecological Ideas*. 2nd ed. Cambridge: Cambridge University Press, 1994.

CHAPTER 5

"Eden's Bad Boy"
Humans and the Animal World in Melville's Poetry

BRIAN YOTHERS

The most unusual animals to appear in Herman Melville's poetry are typically not identified as animals at all. In his late poem "Venice," Melville finds a parallel to human architecture in the work of the animals that create and compose coral reefs:

> *With Pantheist energy of will*
> *The little craftsman of the Coral Sea*
> *Strenuous in the blue abyss,*
> *Up-builds his marvellous gallery*
> *And long arcade,*
> *Erections freaked with many a fringe*
> *Of marble garlandry,*
> *Evincing what a worm can do. (Published Poems 291)*

In this poem, which as Hershel Parker has pointed out, was likely among Melville's earliest poetry to be written, even though it was published only at the end of his life (418), Melville considers a moderately obscure piece of knowledge, that coral reefs, rather than being simply rock or vegetable matter are formed by small and unusual animals known as coral polyps. With this in mind, Melville imagines the polyps collectively as artists driven by a "Pantheist energy of will" to create patterns that humans experience as extraordinarily beautiful and complex when they observe a coral reef. The collective will of the myriads of coral polyps is imagined as an artist's design, and Melville's dry understatement "Evincing what a worm can do" ends the opening stanza of the poem by suggesting that animals can set a remarkable standard for beauty and artistry that humans will struggle to match. The polyp is described as a "craftsman" who builds a "gallery / And long arcade," showing what would appear to be artistic intention, as the "Erections" are "freaked" with "garlandry." The passage acknowledges how, although they cannot be described as expressing conscious intentionality like human artists and architects, the coral polyps create images to which we can respond as if they are self-conscious instances of artistic design (fig. 5.1). Artistry is here attributed neither to a cosmic designer nor to human agency but to the unconscious and instinctual movements of animal life.

The comparison that Melville makes with an unnamed Venetian architect provides both a rebuke to human arrogance and a meditation on how nature can provide the appearance of intentionality in a way that calls into question the idea of human free will:

> *Laborious in a shallower wave,*
> *Advanced in kindred art,*
> *A prouder agent proved Pan's might*
> *When Venice rose in reefs of palaces. (Published Poems 291)*

Here the artistry of the coral polyp becomes a more self-conscious version of a universal will to create. Humans may be "prouder agent[s]" than coral polyps, but they are still controlled by the same underlying forces, and so prove "Pan's might." The title of the poem is suggestive of the deep admiration that Melville expressed throughout his career for Italian art and architecture, but the placement of the coral polyps at the start of the poem is suggestive of Melville's frequently expressed view that the human capacity for meaning making must be read in relation to that of other living things. Michael Jonik aptly describes the poem as "an ode to the creative power of non-human life," a description that could justly be applied to much of Melville's poetry (220). As Colin Dayan has observed, "In his similes, in his metaphors, Melville sought to show the thing-likeness of what might seem human" (48). This characteristic of Melville's work is thrown into high relief in "Venice."

Starting with coral reefs illustrates the extraordinary range of ways in which Melville engages with the animal world through his poetry. Melville's

Figure 5.1. *Brain Coral*. Photo by Jan Derk. Wikimedia Commons (CC BY-A).

most characteristic way of invoking the animal world in his verse is to use a highly specific reference to birds, insects, fish, coral reefs, or seals, among others, to puncture humanity's pretension to cosmic centrality. This becomes particularly evident in poems that compare human structures to natural structures built by nonhuman beings, as in "Venice," or that narrate or suggest a catastrophe like a shipwreck, as in "The Haglets" or "The Berg." Collectively, these poems illustrate just how misleading the narrative of a "long quietus" post-*Moby-Dick*, created by Raymond Weaver, is as an account of Melville's career and how richly his poems interact with his leading themes in his early novels. Martin Kevorkian, Peter Riley, and Sanford Marovitz have made strong cases for the value of Melville's poetry in recent decades, building on earlier work by William Shurr and William Bysshe Stein. Melville's use of animals in his poetry underscores the continuity in his work.

Melville's poetry also evinces a fascination with aspects of the natural world that contrast sharply with the sublimity of the whales about which he is most famous for writing. Although his poems "The Maldive Shark" and "The Man-of-War Hawk" from *John Marr and Other Sailors* harken back to Melville's representations of sharks and birds of prey in *Moby-Dick*, he takes a distinctive approach in "The Maldive Shark" by focusing more on the pilot fish than on the "pale ravener of horrible meat" of the title. His references to the "pintado and gannet" in the "Pebbles" sequence connect his earlier references to seabirds like "Mother Carey's Chicken" and the Catskill eagle from his North American experience on land, which are distinguished by their ability to dive in and out of watery and mountainous gorges alike, with an increasing interest in the ability of birds to move with the winds that might buffet them rather than be overwhelmed by them. Melville's posthumously published collection *Weeds and Wildings* offers a series of reflections on birds and butterflies in which he favorably compares the resilience and buoyancy of birds and winged insects to that of earthbound humans. Perhaps most representative of these moments is a little-known poem titled "Butterfly Ditty," where he writes in the voice of butterflies in a way that brings to mind Emily Dickinson's lyrics: "We'll rove and we'll revel, / Concerned but for this / That Man, Eden's bad boy / Partakes not the bliss" (*Billy Budd* 85). Whether writing about pilot fish, or birds, or butterflies, or, as in one late poem, kittens, by the end of his career Melville had turned from writing about the leviathanic sublime to considering the vitality and beauty of small and adaptable creatures. This chapter anatomizes Melville's transition from the whales of his prose to the butterflies of his poetry.

Animals in Melville's Poems of the Sea

The continuities between Melville's poetry and his better-known representations of animals in *Moby-Dick* are most clear in his poems of the sea, where the elemental power of the sea and the relative helplessness of individual human beings are foregrounded. These appear particularly in the collection *John Marr and Other Sailors*, but also in *Battle-Pieces*, *Clarel*, *Timoleon*, and some of his posthumously published poetry.

"The Haglets," a comparatively long poem that narrates a naval disaster, is the most substantial poem in *John Marr and Other Sailors* where animal imagery plays a major role. This poem narrates the story of an admiral who has been successful in battle but who ultimately will lose his life to the superior power of the sea. The haglets of the title are both witnesses to the calamity narrated and beings that provide a counter to, even a judgment on, human presumption. Haglets, as William Shurr first pointed out, are birds of the shearwater (*Puffinus*) genus, and the fact that Melville uses the colloquial "haglets" for the shearwater associates the birds with the supernatural and witchcraft in particular (132). Robert D. Madison makes the identification still more precise by using Edward Howe Forbush's *Birds of Massachusetts and Other New England States* (1929) to establish that a haglet is specifically the great shearwater (*P. gravis*), a large and graceful seabird (Madison 80). As Shurr suggests (132), Melville introduces the haglets in a way that recalls the appearance of the weird sisters in *Macbeth*:

> *here, peaked and gray, three haglets fly,*
> *And follow, follow fast in wake*
> *Where slides the cabin-lustre shy,*
> *And sharks from man a glamour take,*
> *Seething along the line of light*
> *In lane that endless rules the war-ship's flight. (Published Poems*
>   *219–20)*

Melville presents his readers with the uncanny image of the three Shakespearean haglets alongside a supplementary image that is more familiar to readers of his fiction: that of predatory sharks. The sharks take a "glamour" from human beings because the artificial light provided by the ship illuminates them as they follow in the ship's wake. They also are a reminder of how precarious the position of human sailors is, as the sharks wait to devour anyone who loses the protection of the ship, either through shipwreck or by being thrown overboard. This risk is one that appears repeatedly in Melville's fiction, particularly in *Moby-Dick*, were Ishmael points out that sharks are "the invariable outriders of all slave ships crossing the Atlantic, systematically trotting alongside, to be handy in case a parcel is to be carried anywhere, or a dead slave to be decently buried." There is a gothic dimension to how Melville crafts this scene in bringing together the "diabolism" of the sharks that he has identified in *Moby-Dick* with the foreboding presence of the haglets.

Melville emphasizes the inscrutability of the haglets when he depicts their route as they follow the doomed ship:

> *The sea-fowl here, whose hearts none know,*
> *They followed late the flag-ship quelled,*
> *(As now the victor one) and long*
> *Above her gurgling grave, shrill held*

> *With screams their wheeling rites—then sped*
> *Direct in silence where the victor led. (Published Poems 220)*

The clause "whose hearts none know" points toward the limits of human knowledge of animals and their intentions, but Melville also makes clear that like the coral polyps, the haglets provide apparent evidence of intentions that humans feel compelled to interpret. They seem to be carrying out "rites" of ill omen for the sailors above the victorious ship, which Melville makes increasingly clear is likely to share the fate of the ship that it vanquished.

> *Now winds less fleet, but fairer, blow,*
> *A ripple laps the coppered side,*
> *While phosphor sparks make ocean gleam,*
> *Like camps lit up in triumph wide;*
> *With lights and tinkling cymbals meet*
> *Acclaiming seas the advancing conqueror greet. (Published Poems 220)*

Triumph in a battle at sea has proved to be illusory. Humans may be able to defeat each other, but the power of fate itself is beyond their ability to control, and the "acclaim" offered by the sea is, ironically, the same as the fate of the ship that has been defeated.

In the middle of the poem, Melville again offers the haglets as a sign of the ship's destruction:

> *And, shrilling round, the inscrutable haglets flew.*
> *And still they fly, nor now they cry,*
> *But constant fan a second wake,*
> *Unflagging pinions ply and ply,*
> *Abreast their course intent they take;*
> *Their silence marks a stable mood,*
> *They patient keep their eager neighborhood. (Published Poems 221)*

The haglets continue to offer signs of the natural world's aloofness toward human hopes and fears. On the one hand, the poem projects evil onto the birds; on the other, it is clear that the birds' agenda is not really associated with the terrors that the humans are experiencing. Melville makes clear that, as ominous as they may appear, the haglets have no particular concern with the fate of the ship:

> *Less heeds the shadowing three that play*
> *And follow, follow fast in wake,*
> *Untiring wing and lidless eye—*
> *Abreast their course intent they take;*
> *Or sigh or sing, they hold for good*
> *The unvarying flight and fixed inveterate mood. (Published Poems 223)*

The haglets continue to serve as inscrutable observers to a calamity that is not of their making but to which they appear utterly indifferent.

Finally, the haglets bear witness to the destruction of the ship and the death of the crew:

> *Man keeps from man the stifled moan;*
> *They shouldering stand, yet each in heart how lone.*
> *Some heaven invoke; but rings of reefs*
> *Prayer and despair alike deride*
> *In dance of breakers forked or peaked,*
> *Pale maniacs of the maddened tide;*
> *While, strenuous yet some end to earn,*
> *The haglets spin, though now no more astern.*
> *Like shuttles hurrying in the looms*
> *Aloft through rigging frayed they ply—*
> *Cross and recross—weave and inweave,*
> *Then lock the web with clinching cry*
> *Over the seas on seas that clasp*
> *The weltering wreck where gurgling ends the gasp. (Published*
> *Poems 224)*

The pattern of the haglets' flight now resembles the interweaving of the loom of the three Fates (fig. 5.2). The haglets are compared to the spinners in Greek mythology in a way that directly echoes Ishmael's reflections on Fate, Free Will, and Chance in the "Mat-Maker" chapter of *Moby-Dick*. The haglets have signaled the ship's doom from the start of the poem, and now they are present to observe its culmination.

An irony of Melville's representation of the sublimity of the natural world in "The Haglets" and "The Berg" (discussed later) is of course the fact that in our own century, the capacity of humans to alter or destroy the most majestic aspects of the natural world has come into stark relief, nowhere more dramatically than in the rapid disintegration of glaciers in polar regions. And yet Melville never quite denies the capacity of human beings to damage the world. Rather, he shows the relationship between human fragility and the capacity of humans to destroy and considers the ways in which these two tendencies are complementary.

As the image of the great shearwater included here shows, in addition to being associated with nature's indifference and human foreboding in Melville's poem, it is also a graceful and beautiful bird (something Madison and the ornithologists he cites in "Melville's Haglets" also point out). Melville's interest in the shearwater seems to be functioning on two levels: on one, humans attribute to the birds their own fear of a precarious situation; on another, we are able to see that the birds themselves have no particular need to be defined by human needs. The fact that these are not vultures but birds that many observers might under other circumstances describe as beautiful only heightens the effect. One of the earlier responses to "The

Figure 5.2. Greater shearwater (*Ardenna gravis*) in flight off Sagres, Portugal. Photo by Charles J. Sharp Photography. Wikimedia Commons (CC BY-SA 4.0).

Haglets," William Bysshe Stein's 1958 essay "The Old Man and the Triple Goddess," offers a surprisingly ecological reading of the poem. Stein reflects, "In 'The Haglets,' Melville shows that death is transformation, the outer reflex or manifestation of life energy in the compound of the individual. Hence it becomes, in the cosmic sense, a transfiguration, a reflection of the reciprocal play of night and day, of decay and growth" (47). "The Haglets," for Stein, offers not just the sort of sense of natural indifference to human desires that we might expect to see in Stephen Crane but also a sense of humanity's immersion into the natural world, and the presence of the birds in the poem offers a concrete sense of the place of humans in relation to their surroundings.

Melville extends this sense of humans' location in the natural world when he expresses the sublimity of birds of prey in "The Man-of-War Hawk," where he writes about an especially impressive raptor that draws its name from its association with human beings. Melville describes the bird's beauty and power in the following lines:

> *Yon black man-of-war-hawk that wheels in the light*
> *O'er the black ship's white sky-s'l, sunned cloud to the sight,*
> *Have we low-flyers wings to ascend to his height?*
> *No arrow can reach him; nor thought can attain*
> *To the placid supreme in the sweep of his reign.* (Published Poems 230)

The man-of-war hawk is not described in the sinister terms associated with the haglets. Rather, its power and its transcendence of human concerns come to the fore. The hawk is seen at a great distance, black against the light of the sun, and he defies any attempt to define him in human terms. We are "low-flyers" who can only aspire vainly to the sort of supremacy that he can display in the skies. As in "The Haglets," this passage decenters humans vis-à-vis other animals.

Sharks and Pilot Fish in *John Marr and Other Sailors*

Perhaps Melville's most haunting poem devoted to the animal world is "The Maldive Shark," which captures the relationship between power and beauty in a way that still can disturb us over 135 years after its initial 1888 publication in *John Marr and Other Sailors*. Sharks had been an important element in many of Melville's earlier works, often standing in for the amorality of power, as when he describes their "pantheistic vitality" and "jewel-hilted mouths" and their role as "outriders to slave ships" in *Moby-Dick* (293). In *Clarel*, Melville used the shark as a means to raise the question of theodicy, when a character muses, addressing God, "The shark thou mad'st / Yet claim the dove" (42). In "The Maldive Shark," the shark's power reads as almost oppressive in its passivity, reversing the image of "pantheistic vitality" that appears in *Moby-Dick*. Melville begins the poem, which I quote in full below, by referring to the shark as an oddly sluggish predator, describing it as "phlegmatical" and a "sot," marked less by vitality than by formidable potential:

> *About the Shark, phlegmatical one,*
> *Pale sot of the Maldive sea,*
> *The sleek little pilot-fish, azure and slim,*
> *How alert in attendance be.*
> *From his saw-pit of mouth, from his charnel of maw*
> *They have nothing of harm to dread,*
> *But liquidly glide on his ghastly flank*
> *Or before his Gorgonian head;*
> *Or lurk in the port of serrated teeth*
> *In white triple tiers of glittering gates,*
> *And there find a haven when peril's abroad,*
> *An asylum in jaws of the Fates!*
> *They are friends; and friendly they guide him to prey,*
> *Yet never partake of the treat—*
> *Eyes and brains to the dotard lethargic and dull,*
> *Pale ravener of horrible meat.* (Published Poems 236)

The poem ends as it began, with the idea that the shark is both terrifying and strangely inert. Its mouth appears as a "port of serrated teeth / In white triple tiers of glittering gates"; it remains a "dotard lethargic and dull." The final line makes clear that the shark asserts its power when it has the opportunity to eat, becoming a "Pale ravener of horrible meat," but through the bulk of the poem the shark's distinguishing quality is that its power is potential rather than active. The passivity of the shark's power contrasts sharply with the activity of the pilot fish, which are as harmless in themselves as the shark is horrifying, but they also play a role in a world where predation is inescapable. It is noteworthy that both the pilot fish and the shark are described in terms that humanize them. The sympathetically portrayed pilot fish appear in the role of alert servants who need to adapt their conduct to the needs of their masters. The shark is described in terms associated with the gluttony and drunkenness of the idle and imperious rich.

While Melville captures the pilot fish's mutually beneficial relationship to the shark, he overstates their role when he suggests that the pilot fish "guide" the shark to its food; rather, they accompany the shark and share in the food from its kills (fig. 5.3). The poem is one of Melville's few works about nature that is not as much interested in setting up a contrast between the natural world and the world of human beings as it is in examining a natural phenomenon, and here the contrast is between two animals: the fearsome shark and the harmless pilot fish that depend on the shark. Robert Pinsky has identified "an appealingly aggressive, tough quality to the poem, almost as though the poet is thinking about more sentimental, cloying approaches to this same material, such as symbiosis, or the grace of fish." Pinsky is right to note Melville's lack of sentimentality in the poem. Particularly important here is the contrast between the shark's terrifying power and amorality and its companions' appealing ability to live in the mouth of the shark without being devoured, coupled with the idea that the pilot fish are dependent in some way on the violence embodied by the shark. Douglas Robillard has pointed out that Shakespeare provides a likely source for Melville's depiction of the shark in this poem, noting that the use of "ravined" to describe sharks in *Macbeth* seems a likely source for Melville's "pale ravener" (80). As in "The Haglets," Melville's use of Shakespeare serves both to attribute human characteristics to nonhuman beings and to undermine the idea of human centrality to the natural world.

Another of Melville's most haunting poems of the sea in *John Marr and Other Sailors* is one that focuses less on individual animals than on an environment that can nourish them, in this case, somewhat implausibly, an iceberg. In "The Berg," Melville shows, as Marissa Grunes has recently pointed out in an online piece that features a video interview with glaciologist Catherine Walker, that animals can flourish in a place that can be a site of terror for human beings. Melville emphasizes the threat that an iceberg can pose to ships at sea, conveying a sense of dread through his description of the sheer size and sublimity of the iceberg:

Figure 5.3. Shark (*Carcharhinus longimanus*) with pilot fish (*Naucrates ductor*) at the Elphinstone Reef in the Red Sea off Egypt. Photo by Peter Koebl. Wikimedia Commons (CC BY-SA 2.5).

Along the spurs of ridges pale,
Not any slenderest shaft and frail,
A prism over glass-green gorges lone,
Toppled; or lace of traceries fine,
Nor pendant drops in grot or mine
Were jarred, when the stunned ship went down.
Nor sole the gulls in cloud that wheeled
Circling one snow-flanked peak afar,
But nearer fowl the floes that skimmed
And crystal beaches, felt no jar.
No thrill transmitted stirred the lock
Of jack-straw needle-ice at base;
Towers undermined by waves—the block
Atilt impending—kept their place.

> *Seals, dozing sleek on sliddery ledges*
> *Slipt never, when by loftier edges*
> *Through very inertia overthrown,*
> *The impetuous ship in bafflement went down. (Published Poems*
>   *240–41)*

Grunes and Walker have pointed out that Melville here acknowledges the unexpected degree of biodiversity that appears on an iceberg, a site that would appear to be hostile to life in general, but which in fact supports animal life even as it threatens the lives of humans on a ship. As in "The Haglets," humans here are not allowed to have the final word with regard to their own survival, and the animal world is able to maintain a degree of indifference to the fate of individual human beings and the ships on which they sail. A ship may sink as a result of striking an iceberg without dislodging a seal or even impinging upon its nap or disrupting the flights of birds about the iceberg, and human terror may not even register on the other lifeforms that flourish on the iceberg.

From the Sea to the Garden to the Desert: Butterflies, Birds, and Camels in Melville's Posthumously Published Poetry and *Clarel*

Many of Melville's intuitions about the relationship between humans and animals are captured in "Butterfly Ditty." This poem appears as part of *Weeds and Wildings, with a Rose or Two*, which Melville left in manuscript form at the time of his death in 1891. The poem reflects on the state of human beings in relation to a natural world that is here playfully envisioned as a paradise. The butterfly speaker reflects:

> *Summer comes in like a sea,*
> *Wave upon wave how bright;*
> *Thro' the heaven of summer we'll flee*
> *And tipple the light!*
> *From garden to garden,*
> *Such charter have we,*
> *We'll rove and we'll revel,*
> *And idlers we'll be!*
> *We'll rove and we'll revel,*

> *Concerned but for this,—*
> *That Man, Eden's bad boy,*
> *Partakes not the bliss. (Billy Budd 85)*

This poem might seem to be miles away from the "pale ravener of horrible meat," but there is something here of Melville's increasing interest in the contrast between predators that dominate their landscape, like sharks and humans, and creatures like the pilot fish and butterflies that find means of survival through agility rather than dominance and that can seem to find grounds for resilience and even pleasure in the midst of mortality.

The poem is voiced as the words and thoughts of a butterfly, and Melville expresses what is, late in his career, becoming a recurring theme: that pleasure is to be preferred to power, freedom to property, and that humans' unique relation to power and property in the animal world is precisely what keeps them from entering into or remaining in an Edenic state. To some degree, this emphasis takes us back to Melville's first published book, *Typee*, in which he describes nineteenth-century capitalism in the United States and Europe as a falling away from paradise, but here the idea seems to be that there is something essential in human nature, as opposed to merely in North American and European civilization, that drives us out of Eden. The butterflies are imagined at the end of the poem, on one possible reading, as the angels who keep humans from Eden, but the clear implication is that humans' status as "Nature's bad boy" is sufficient to deprive them of bliss. Melville's phrase "concerned but for this" can suggest that butterflies feel compassion for human inability to embrace the Eden around them, but also that they, like the angels with the flaming swords in Genesis, have a responsibility to see that humans do not share their bliss. What is clear is the failure of humans to be able to take the pleasure in the natural world that would be their birthright if they resembled other animals more closely in harmonizing with their surroundings. A major suggestion here is that the fall of humanity is correlated to the human need to control and dominate their surroundings, sometimes to their own detriment.

Melville also makes ironic use of the legal term "charter," as he imagines the butterflies having a kind of ownership over the landscape that humans, with their established legal claims, may aspire to but fail to attain. The use of "charter" here is reminiscent of Thoreau's reflection on landownership in *Walden*, where he writes, "I have frequently seen a poet withdraw, having

Figure 5.4. Illustrations of new species of exotic butterflies selected chiefly from the collections of W. Wilson Saunders and William C. Hewitson, vol. I (London: John van Voorst, 1869). Wikimedia Commons.

enjoyed the most valuable part of a farm, while the crusty farmer supposed he had got a few wild apples only" (86). Like Thoreau, Melville is here imagining the landscape as spiritually rather than materially valuable, and Melville's butterflies play a similar role to that of Thoreau's poet (fig. 5.4).

This poem pairs well with another of Melville's late, unpublished poems,

"Montaigne and His Kitten," which I have discussed elsewhere and features a human speaker who sees a deeper wisdom in the gambols of a kitten. This poem implies that animals may be more likely to merit immortality than their human masters (Yothers 107). There is also an affinity here with Melville's celebrated younger contemporary Emily Dickinson, whose poem "I taste a Liquor never brew'd" also imagines the pollination of plants as a form of drunkenness, or tippling.

Although Melville focuses especially on birds, winged insects, and sea creatures in his poetry, land animals also have a role to play in deflating human presumption. In *Clarel*, Melville's 1876 *Poem and Pilgrimage in the Holy Land*, which ran to eighteen thousand lines, camels serve to humble humans in a manner that is full of Melville's characteristic irony. Nehemiah, a fanatical American Protestant missionary in Palestine, is passing out tracts to anyone who will take them, attempting to persuade Jews and Muslims to convert to Christianity and Catholic and Orthodox Christians to convert to Protestantism. The narrator's attitude toward Nehemiah is frequently marked by irony, as readers of Melville's earlier critiques of Christian missionaries in *Typee* and *Omoo* might suspect. Nehemiah attempts to share his tracts with a passing Muslim, only to be thwarted by his camel:

> *Under gun, lance, and scabbard hacked*
> *Pressed Nehemiah; with ado*
> *High he reached up an Arab tract*
> *From the low ass—"Christ's gift to you!"*
> *With clatter of the steel he bore*
> *The lofty nomad bent him o'er*
> *In grave regard. The camel too*
> *Her crane-like neck swerved round to view;*
> *Nor more to camel than to man*
> *Inscrutable the ciphers ran.*
> *But wonted unto arid cheer,*
> *The beast, misjudging, snapped it up*
> *And would have munched, but let it drop;*
> *Her master, poling down his spear*
> *Transfixed the page and brought it near,*
> *Nor stayed his travel. (Clarel 173)*

The image here is deeply comical, of course. Melville imagines Nehemiah, the evangelical missionary who is attempting to convert Jewish and Muslim inhabitants of Palestine to his version of Christianity, to be passing out tracts, and the camel who snatches the tract declines to eat it, despite initially considering it because accustomed to "arid cheer." Here Melville's critique of human presumption blends with his critique of religious exclusivism, and he uses the camel to satirize the presumptuousness of attempts to convert people from one faith to another without having truly understood the relationship among the Abrahamic faiths. At the same time, the millennium-old rivalries among these faiths in Palestine are irrelevant from the vantage point of the camel, whose only concern is whether the printed matter is worth munching on. Something of Melville's comic use of the camel is captured in an image by J. S. Küslen that Melville kept in his own print collection, as documented in *Melville's Print Collection Online* (Wallace, Otter, and Farrell) in which a camel stares straight ahead while being restrained by his attendant (fig. 5.5).

Elsewhere in *Clarel*, Melville uses birds in ways that parallel his recurrent interest in birds throughout his career. As I have discussed elsewhere, Melville's depiction of Mother Carey's chickens in *Clarel* has affinities with his earlier allusions to the same birds in *Moby-Dick* and also to his celebrated meditation on the Catskill eagle in *Moby-Dick*, and these moments also connect meaningfully with Melville's Civil War poem "Shiloh," which takes the flight of swallows as its framing image. Another moment that highlights what birds mean to Melville is his description of the pigeons in the monastery at Mar Saba, where Melville delves more deeply into the religious implications of the presence of these birds through the reflections of Derwent, an Anglican clergyman:

> *Those pigeons, now, in Saba's hold,*
> *Their wings how winsome would they fold*
> *Alighting at one's feet so soft.*
> *Doves, too, in mosque, I've marked aloft,*
> *At hour of prayer through window come*
> *From trees adjacent, and a'thrill*
> *Perch, coo, and nestle in the dome,*
> *Or fly with green sprig in the bill.*
> *How by the marble fount in court,*

Figure 5.5. *Esclave tenant un chameu par la bride* (Slave holding a camel by the bridle), by J. S. Küslen [after Stefano della Bella]. From della Bella's *Plusieurs têtes coiffées à la persiennes*, 1649. *Melville's Print Collection Online*, edited by Robert K. Wallace, Samuel Otter, and Clementine Farrell. Courtesy of the Melville Chapin Collection, photo by Harvard Library Services.

> *Where for ablution Turks resort*
> *Ere going in to hear the Word,*
> *These small apostles they regard*
> *Which of sweet innocence report.*
> *None stone the dog; caressed, the steed;*
> *Only poor Dobbin (Jew indeed*
> *Of brutes) seems slighted in the East. (Clarel 413)*

Several elements of this meditation are worth stressing: first, the attentive affection with which the pigeons are described; second, the way in which the birds are connected with human acts of worship; and third, the way in which cultural difference is mediated through animals in this instance. Melville describes the way that the birds fold their "winsome" wings and how they "Perch, coo, and nestle in the dome" of a mosque. This leads into how Muslims attending services at the mosque regard the birds: as "small apostles . . . / Which of sweet innocence report." The birds, in other words, offer a message from the divine, and the worshippers recognize this. This reverence for the birds translates into a broader affection for animals in the next line. Melville's description of the pigeons here suggests that one measure he uses in assessing human culture is how it treats animals.

Similarly, here, Melville relates religious bigotry and cruelty to the abuse of animals, referring to the Dobbin, or draft horse, as the "Jew indeed / Of brutes." The bigotry that Melville observes among Christians and Muslims against Jewish people finds an analogy in the mistreatment of a class of animals. It is also noteworthy that there is a subtle class as well as religious dimension: the "steed" (a horse that humans ride) is "caressed," whereas a horse used for more mundane labors "seems slighted."

Melville's commitment to understanding and representing religious difference in *Clarel* has been discussed by numerous critics, and one point that has been made repeatedly is that Melville treats both Islam and Judaism with considerable sympathy and sensitivity in his longest poem. What is especially notable here is the way in which the favorable view of Islam is directly related to Muslims' treatment of animals. Human virtue, Melville seems to suggest, is directly correlated to how humans interact with those with whom they share the planet, and Melville seems moved by the way in which the worshippers at the mosque value the pigeon that flies through it. Melville's treatment of animals, here as in so many other places in his work, is interwoven with his treatment of interrelationships among humans, from his criticisms of racism and imperialism to his explorations of the various approaches religious traditions take to the divine.

Melville's Animals and the Anthropocene

There is a further valence to "Eden's bad boy," the line Melville uses to describe humans in his butterfly poem, and it takes us back to one of the crucial ecological quandaries in the interpretation of Melville's work. As discussed previously, Melville not infrequently seems to place the ability to disrupt nature seriously outside the range of human possibilities, a plausible viewpoint in the nineteenth century that seems significantly less plausible in our own times. A commonly voiced criticism of *Moby-Dick* is that Melville, via Ishmael, downplayed the possibility that humans could hunt whales to the point of extinction, as Ishmael concludes the chapter "Does the Whale's Magnitude Diminish—Will He Perish?" with this reassuring reflection:

> Wherefore, for all these things, we account the whale immortal in his species, however perishable in his individuality. He swam the seas before the continents broke water; he once swam over the site of the Tuileries, and Windsor Castle, and the Kremlin. In Noah's flood he despised Noah's Ark; and if ever the world is to be again flooded, like the Netherlands, to kill off its rats, then the eternal whale will still survive, and rearing upon the topmost crest of the equatorial flood, spout his frothed defiance to the skies. (*Moby-Dick* 462)

The passage seems to revel in human puniness relative to the whale and so to hold out the hope that nothing humans can do can ultimately destroy the sperm whale as a species. This passage has often given pause to critics who wish to understand Melville as an environmental writer, as it seems complacent with regard to the damage that humans can do in relation to the most majestic species in the natural world. Melville's formulation "Eden's bad boy" deserves more attention, especially in relation to this question, and it offers us a way of understanding how Melville might acknowledge human capacity to destroy the natural world even as he criticizes human presumptions of centrality within that world.

Another way of saying "Eden's bad boy," the phrase that gives this chapter its title, might be captured in a poem from the "Pebbles" sequence in *John Marr* that only indirectly refers to the natural world. In the fourth epigram in the sequence, Melville wrote "On ocean where the embattled fleets repair, / Man, suffering inflictor, sails on sufferance there" (*Published Poems* 246). "Eden's bad boy" cannot help being a "suffering inflictor," and the sense that our power in the world is but a matter of "sufferance" indicates that human arrogance is ultimately checked by the limits of human agency. The formulation "suffering inflictor" is suggestive of a more sophisticated environmental consciousness than Melville is sometimes credited with. Yes, nature is bigger than human beings, and Melville has a strong and consistent commitment to writing about the sublimity of nature, but human beings are themselves "inflictors" who can wreak damage on the natural world even as they are incapable of fully comprehending its vastness. The story of Eden provides an example of how human moral failings can destroy a natural world that could otherwise be a source of beauty, sublimity, awe, and indeed bliss.

This pattern of acknowledging both the sublimity of nature and the limits of human knowledge is a central theme in the "Pebbles" sequence that ends *John Marr*. In the first poem in the sequence, Melville writes,

> *Though the Clerk of the Weather insist,*
> *And lay down the weather-law,*
> *Pintado and gannet they wist*
> *That the winds blow whither they list*
> *In tempest or flaw. (Published Poems 243)*

As Wyn Kelley has noted regarding this poem, the "pintado" is a name for a Cape petrel that means "painted" in Portuguese; Kelley points out that Melville's use of this specific name for the bird emphasizes "how vividly the bird's black-and-white plumage must appear in a squall" (134). Tom Nurmi, meanwhile, has shown that the "Pebbles" sequence plays a further environmental role in its blending of ecology and technology, an issue raised by the pintado and gannet in particular. The pintado and gannet have the function of showing how animals can live in a world that humans' advanced knowledge has still not allowed them to master. Indeed, a storm that human experts (at least nineteenth-century ones) can neither predict nor control can be managed by the instincts of the seabirds Melville invokes. Melville's reflection on the pintado and gannet addresses the ways in which humans fall short of other animals in integrating themselves into their environment.

Melville's poetry has frequently been dismissed by critics as lacking the sheer formal exuberance of *Moby-Dick*, but there is a sense in which Melville offers a matured and powerful version of his insights in *Moby-Dick* through this later work. The relationship between the human and animal worlds continues to be fraught, but Melville increasingly finds ways of acknowledging telling details that illuminate the complexities of a human relationship to the animal world that is, or at least can be, based on something other than dominance and exploitation. Melville speaks to our own uncertain moment in that he shows that humans must embrace humility in their relationship to the natural world and that the alternative to such humility is catastrophic failure.

Works Cited

Dayan, Colin. "Melville's Creatures, or Seeing Otherwise." In *American Impersonal: Essays with Sharon Cameron*, edited by Branka Arsic, 45–56. New York: Bloomsbury, 2014.

Grunes, Marissa. "What Happens When Icebergs Collide with Art." *Nautilus*, November 2, 2022. https://nautil.us/what-happens-when-icebergs-collide-with-art-244877/.

Jonik, Michael. *Herman Melville and the Politics of the Inhuman*. New York: Cambridge University Press, 2018.

Kelley, Wyn. "Lauding the Inhuman Sea." *Leviathan: A Journal of Melville Studies* 17, no. 1 (2015): 133–35.

Kevorkian, Martin. "Faith among the Weeds: Melville's Religious Wildings beyond These Deserts." In *Visionary of the Word: Melville and Religion*, edited by Jonathan A. Cook and Brian Yothers, 97–128. Evanston, IL: Northwestern University Press, 2017.

Madison, Robert D. "Melville's Haglets." *Leviathan: A Journal of Melville Studies* 5, no. 2 (2003): 79–83.

Marovitz, Sanford E. *Melville as Poet: The Art of "Pulsed Life."* Kent, OH: Kent State University Press, 2013.

Melville, Herman. *Billy Budd, Sailor, and Other Uncompleted Writings*. Edited by Harrison Hayford, Alma A. MacDougall, Robert A. Sandberg, G. Thomas Tanselle, and Hershel Parker. Evanston and Chicago: Northwestern University Press and The Newberry Library, 2019.

———. *Clarel, A Poem and Pilgrimage in the Holy Land*. Edited by Harrison Hayford, Alma A. MacDougall, Hershel Parker, and G. Thomas Tanselle. Evanston and Chicago: Northwestern University Press and The Newberry Library, 1991.

———. *Moby-Dick, or, The Whale*. Edited by Harrison Hayford, Hershel Parker, and G. Thomas Tanselle. Evanston and Chicago: Northwestern University Press and The Newberry Library, 1988.

———. *Published Poems*. Edited by Robert C. Ryan, Harrison Hayford, Alma MacDougall Reising, and G. Thomas Tanselle. Evanston and Chicago: Northwestern University Press and The Newberry Library, 2009.

Nurmi, Tom. *Magnificent Decay: Melville and Ecology*. Charlottesville: University of Virginia Press, 2020.

Parker, Hershel. *Herman Melville: A Biography*. 2 vols. Baltimore: Johns Hopkins University Press, 1996 (vol. 1) and 2002 (vol. 2).

Pinsky, Robert. "Truth in Darkness." *Slate*, January 24, 2012. http://www.slate.com/articles/arts/classic_poems/2012/01/herman_melville_s_the_maldive_shark_.html.

Riley, Peter. *Whitman, Melville, Crane, and the Labors of American Poetry: Against Vocation*. Oxford: Oxford University Press, 2019.

Robillard, Douglas. "Melville's 'Pale Ravener of Horrible Meat.'" *Leviathan: A Journal of Melville Studies* 8, no. 2 (2006): 85.

Shurr, William. *The Mystery of Iniquity: Melville as Poet, 1857–1891*. Lexington: University Press of Kentucky, 1972.

Stein, William Bysshe. "The Old Man and the Triple Goddess: Melville's 'The Haglets.'" *Journal of English Literary History* 25 (1958): 43–59.

Thoreau, Henry David. *The Illustrated Walden: The Thoreau Bicentennial Edition*. New York: Penguin Tarcher Perigree, 2016.

Wallace, Robert K., Samuel Otter, and Clementine Farrell, eds. *Melville's Print Collection Online: A Pictorial Fusion of His Mind and Vision*. http://melvillesprintcollection.org/exhibits/show/project-and-site/project-and-site-intro.

Weaver, Raymond. *Herman Melville: Mariner and Mystic*. 1920. Reprint, New York: Cooper Square, 1961.

Yothers, Brian. "Whales, Mother Carey's Chickens, and a Heart-Stricken Moose." In *Animals in the American Classics: How Natural History Inspired Great Fiction*, edited by John Cullen Gruesser, 88–110. College Station: Texas A&M University Press, 2022.

CHAPTER 6

"Versed in Country Things"
Animals in the Poetry of Robert Frost

PHILIP EDWARD PHILLIPS

For SJP

Although born in San Francisco, California, Robert Frost (1874–1963) is widely associated with the life and landscape of rural New England, a region that the poet knew well from his youth as a student, a schoolteacher, and a poultry farmer. Frost universalized rural New England in his lyric verse from his first published collection, *A Boy's Will* (1913), to his final collection, *In the Clearing* (1962). As Virginia F. Smith observes in *A Scientific Companion to Robert Frost*, Frost was "a formidable observer of nature" who included over one hundred species of plants (including flowers, berries, grapes, grasses, trees, and other plants) and slightly more species of animals that "fly, crawl, swim, hop, and run" in his verse (2–3).[1] No geographic place has a greater claim to Frost than his farm in Derry, New Hampshire, where he and his young family lived from 1900 to 1911 and where Frost developed a deep appreciation for the natural world and a curiosity about its inhabitants.

In *Can Poetry Save the Earth? A Field Guide to Nature Poems*, John Felstiner affirms Frost's connection to New England and its agrarian mode of life, noting that Frost "look[ed] back on the Derry years as a sacred source of his vocation" (118). The decade when Robert and Elinor Frost lived and raised their family on the thirty-acre farm in Derry was by all accounts a formative and imaginatively fertile time for the poet (fig. 6.1). On the farm, Frost immersed himself both practically and poetically in "country things," with some of his attitudes influenced by Ralph Waldo Emerson's *Nature* and Henry David Thoreau's *Walden*. Although Frost did not publish his first book of poetry until moving to England (where he and his family lived from

1912 to 1915), he told Richard Poirier in an interview for the *Paris Review* that he had already "written three books—*A Boy's Will*, *North of Boston*, and part of the next (*Mountain Interval*) in a loose-leaf heap" that he brought with him (92). The poems in these collections, many of them among his best, convey the rhythms, sounds, and complexity of nature and its human and nonhuman animal inhabitants.

Frost's relationship with nature can be expressed well by the word "curiosity." In *Robert Frost: A Life*, American poet and biographer Jay Parini uses this word when accounting for Frost's ability to bring "[his New England] region to life with [such] unusual specificity" (447). Parini argues that Frost, like William Faulkner, "understood that a literary artist must inhabit a specific place and learn the speech of that place and time" (447). Indeed, Frost's curiosity resulted in his impressive absorption and adept use of local speech patterns, as exemplified by the dialogues in his groundbreaking narrative poetry. In respect to nature, Frost's curiosity led to a prodigious knowledge of the local "birdcalls, the names of flowers, [and] their patterns of blooming" (447), details that his daughter Lesley later recalled having learned to appreciate from her father. Although Frost often uses the natural world

Figure 6.1. Robert Frost Farm, Derry, New Hampshire, July 2023. Photograph by author.

(including its creatures) as "a source" of poetic metaphors and imagery, it nevertheless remains "real." That Frost could "keep the figurative and the literal in balance," Parini argues, was "his genius" (447–48), or at least a major part of it.

In his public persona and in his poetry, Frost cultivated the image of himself as a folksy sage, one who lived close to the land and captured the language of the common person in his poetry (fig. 6.2). As Frost's career advanced, literary critics noticed another side to the poet. In *Poetry and the Age*, American poet and critic Randall Jarrell distinguished between the Frost whom "everybody knows" and the other Frost whom "no one even talks about" (36). The former is "a sensible, tender, humorous poet who knows all about trees and farms and folks in New England" and whose poetry is "fairly optimistic," not "too hard or odd or gloomy," and generally "reassuring" (36). The Frost who "matters most" to Jarrell, however, is the one whose poems are "extraordinarily subtle and strange" (37). This chapter examines a selection of animal poems that in most cases reveal both "Frosts" to us even as they ask hard questions about humanity's (and nonhuman animals') place in the natural world and, in several cases, use nonhuman (and human) animals metaphorically to examine the complexities and potential of the poetic craft.

It is appropriate to begin a discussion of animals in Frost's poetry with "The Pasture," which first appeared as the prologue to *North of Boston*, initially published in London in 1914. Frost's contemporary Amy Lowell noted in her *Tendencies in Modern American Poetry* that this "little poem," originally printed in italics, could "very well serve as motto to all Mr. Frost's work" (104–5). Indeed, beginning with *Collected Poems* (1930), Frost used "The Pasture" as such in all subsequent collections of his verse. Lowell catches the spirit of the poem: "Here in a few words is an upland pasture with the farmer at work in it, and here is the tenderness, that love of place and people which marks all that this poet does" (105). It can be read as a programmatic poem in the classical sense, whose subject, style, and language introduce readers to the volume's central thematic concerns and aesthetic qualities. Reminiscent of the pastoral poetry of Theocritus and Vergil, it is also a *vade mecum* inviting the reader to follow him and to experience the bucolic world represented in *North of Boston* and other volumes.

In the first stanza, the speaker states his intention to "clean the pasture spring," "to rake the leaves away," and, perhaps, to "wait to watch the water

Figure 6.2. Robert Frost, ca. 1955. Photograph by Clara Sipprell. National Portrait Gallery, Smithsonian Institution, bequest of Phyllis Fenner.

clear" (*CPP&P* 13).[2] In the second, the speaker announces that he is "going out to fetch the little calf / That's standing by its mother," creating a nurturing, maternal image and a youthful, precarious one for the calf, who "totters when she licks it with her tongue." The refrain of the first stanza is repeated as the refrain of the second: "I sha'n't be gone long.—You come too." His companion could refer "with a hopeful gesture toward Elinor," as Felstiner suggests (118), and likely did, according to Lesley Frost, who recalled from childhood her father calling to her mother, who was washing dishes inside, to join him outside. The speaker is also inviting the reader to witness the poet at work clearing the spring and fetching the calf or, rather, making a new beginning in poetry that draws inspiration from the spring. Regardless of its intended addressee(s), the poem's tone is inviting and its language down to earth, suggesting that what the speaker has to say here, and throughout the collection, will be relatable to all readers. That having been said, Frost asserted in "Education by Poetry" (and elsewhere on numerous occasions) that poetry "provides the one permissible way of saying one thing and meaning another" (*CPP&P* 719), an approach akin to Emily Dickinson's "Tell[ing] all the truth but tell[ing] it slant."[3] This applies even to a seemingly simple poem about a "little calf" (to which we will return later) as well as to more complex ones.

In *Robert Frost and the Challenge of Darwin*, Robert Faggen argues that "many of Frost's most beguiling poems are about other creatures, especially birds and insects," and that these poems often "explore [humanity's] anxiety about [its] relationship to the rest of the creaturely world" (53). His poems usually exceed the Romantic tendency to anthropomorphize animals, which also become "visionary symbols" reminiscent of the medieval tradition of "animal fable." By invoking this tradition, Faggen argues, "Frost leads us to expect revelation of the transcendent" when "the power of these poems" actually "lurks in their subversion of this expectation" (53). Indeed, Frost's poetry often implies a separation between human and nonhuman animals. However, the poet frequently blurs those boundaries, questioning "anthropocentric assumptions and Judeo-Christian myths of redemption" and suggesting perhaps a post-Darwinian world of "creaturely struggle" (53–54). Refusing to take a clear position, Frost instead revels in suggestiveness and ambiguity.

Like Frost himself, the speakers in his poems are curious about animals. But his human speakers do not hold up a mirror to nature. Rather, the speakers' representations of nonhuman animals in the natural world reflect their own perceptions and limitations. In the three sections that follow, we consider poems about human and nonhuman animal interactions, poems about animals that speak without speaking, and poems about human (mis)readings of nonhuman animals and the natural world. These poems include some of Frost's most memorable birds, horses, cattle, frogs, and insects, as well as human speakers, who observe, reflect on, and engage with their fellow creatures.

Human and Nonhuman Animal Interactions

Frost's poems depicting encounters between human and nonhuman animals usually present the latter as playing a supportive rather than a central role. Even so, their seeming unimportance in "Stopping by Woods on a Snowy Evening," "The Wood-Pile," and "The Exposed Nest" can disguise an underlying significance. Readers of Frost are perhaps most familiar with the "little horse" in "Stopping by Woods," who plays such a role. In this poem from *New Hampshire* (1923),[4] the speaker projects himself onto his horse, who "must think it queer" that he has stopped "without a farmhouse near / Between the woods and frozen lake / The darkest evening of the year" (*CPP&P* 207), a journey reminiscent of Dante's *Commedia*.[5] The speaker remarks that his horse, in response to this irregularity, "gives his harness bells a shake / To ask if there is some mistake." But the speaker, undeterred by the sound of the bells, or "the sweep / Of easy wind and downy flake," remains fixated on the "woods," which are "lovely, dark and deep" (207) and seems determined to listen to their call.

In *Robert Frost: The Work of Knowing*, Richard Poirier regards the depth and darkness of the woods to be "ominous," and he notes "a furtive impulse toward extinction" in the speaker, but one "no more predominant in Frost than it is in nature" (181). The speaker's choice *not* to succumb to this impulse may owe something to his little horse, who reminds him that he has "promises to keep, / And miles to go before I sleep, / And miles to go before I sleep" (*CPP&P* 207). Thus, the little horse's reaction to his unusual behavior tells us more about the speaker than his equine companion. Frost's ulteriority emerges in the repeated lines that conclude this otherwise regular, fourteen-line English sonnet, and draw attention to a kind of "sleep" that seems so attractive to but is ultimately rejected by the speaker thanks to the intervention of his horse.

"The Wood-Pile," a poem from Frost's earlier collection, *North of Boston*, features a speaker on another kind of Dantean journey. "Out walking in the frozen swamp one gray day," the speaker must choose whether to "turn back" or to "go on farther" (*CPP&P* 100). Disoriented, the speaker cannot "say for certain I was here / Or somewhere else: / I was just far from home." At length, he encounters "A small bird," who instinctively keeps his distance, but whose actions the speaker, according to Onno Oerlemans, "finds surprisingly easy to interpret" (135). Indeed, Frost's speaker first observes, "He [the bird] was careful / To put a tree between us when he lighted, / And say no word to tell me who he was," before showing some self-awareness in regarding himself as "foolish . . . to think what *he* thought" (*CPP&P* 100). Nevertheless, as Oerlemans observes, the speaker remains "more interested in himself" and his own thoughts than those of the bird (136). This human-bird encounter becomes an occasion for the poet to highlight the difficulties of one creature really understanding the other. It also suggests a human "need to find agency and purpose in the life of another creature" (137) as well as to attempt to find answers to our (human) questions in the actions or reactions of (nonhuman) animals.

As suddenly as he had shifted his attention from the "frozen swamp" and the similarities of the "tall slim trees / Too much alike to mark or name a place by," the speaker forgets the bird, "let[ting] his little fear" that he had foolishly attributed to the bird "Carry him off the way I might have gone, / Without so much as wishing him good-night" (*CPP&P* 100). Thereafter, he turns his attention to a "pile of wood," something more familiar to him. Unlike this bird, the wood is identified as "a cord of maple." It is the product of human labor, carefully "cut and split / And piled—and measured, four by four by eight," recalling how poetry is similarly measured. The wood is "older sure than this year's cutting," suggesting that someone's labor had been expended for nothing. Seeing that nature has bound it with "Clematis," the speaker is perplexed that "Someone who lived in turning to fresh tasks / Could so forget his handiwork" (101). In *The Art of Robert Frost*, literary critic Tim Kendall suggests the speaker's reaction is one of disapproval, and yet he is one who himself has left "the world of fireplaces" and does not acknowledge his own "forgetfulness," having "dismiss[ed] the bird with the abrupt phrase, 'I forgot him'" (172). Frost's speaker reveals his own tendency to read the world, both natural and animal, according to his own abilities and limitations.

Finally, "The Exposed Nest," from *Mountain Interval* (1916), features a speaker and an unnamed companion who encounter a nest of small birds left exposed by mowing. At the heart of the poem is the human speaker's struggle to decide what, if anything, to do with the helpless birds and the implied implications of that decision. Frost would have been familiar with this kind of situation because of his firsthand experience mowing the grass in the field behind his farmhouse. The speaker and his companion show concern for the "defenseless" birds. Their instinct is to protect them from the "heat and light" and to "restore them to their right / And too much world at once—could means be found" (*CPP&P* 107). Their compassion for the young birds resembles that of the absent mother bird (who may not return because of human "meddling")—toward them. The desire to restore the birds "to their right" and to shield them from the harshness of the world shifts the attention from avian to human concerns. The speaker wants to make things "right" for the birds, but the categories of "right" and "wrong" do not apply here. The poem confronts the reality that all life (human and nonhuman animal) must go on and concludes with the speaker admitting that he has no memory "of ever coming to the place again / To see if the birds lived the first night through, / And so at last to learn to use their wings" (107). What the human protagonists fail to do (or choose not to do) here, perhaps the poet can do by poetically imagining this moment of human and nonhuman animal interaction.

In all three poems, Frost highlights the complexities—and the surprises—that may arise from human and nonhuman animal interactions. The animals represented in these poems play important roles not as the primary subjects of the poet's examination but rather as a means by which the human speakers question their life decisions, reflect on their perceptions of the natural world, and consider their responsibility as nature's stewards or as fellow creatures.

Animals That Speak without Speaking

In the previous section, we discussed animals whose specific identities are less important than their supporting roles. In this section, we consider specific animals in two poems whose names are included in their respective titles. Although they are not solely poems about these animals per se, Frost's use of them has important implications in respect to his assertion of his unique and distinctively American, poetic voice. "Hyla Brook" and "The Oven Bird," which appear back to back in *Mountain Interval*, illustrate Frost's interest in the distinguishing qualities of these animals as well as their metaphorical potential. These animals (or what they represent) allow the poet to test the limits of metaphor to express the poetic voice and to interrogate the poetic craft. Like many other poets,[6] Frost looks to animals capable of song (or speech-like song) to illustrate the capabilities of poetry and the work of the poet (and to distinguish himself and his poetry from others').[7]

"Hyla Brook," an irregular sonnet of fifteen lines, takes its name from an intermittent stream on the Derry farm that flows during the spring but dries up in the summer (fig. 6.3). Frost named the brook after the "peepers," tiny tree frogs of the genus *Hyla* (named after Hylas, the young companion of Hercules abducted by water nymphs) that loudly heralded the coming of spring. The nostalgic poem opens in June, when the speaker states, "our brook's run out of song and speed," either "to have gone groping underground" and "taken with it all the Hyla breed" or "flourished and come up in jewel-weed / Weak foliage that is blown upon and bent / Even against the ways its waters went" (*CPP&P* 115–16).[8] Although the speaker describes the hylas as having "shouted in the mist a month ago, / Like ghosts of sleigh-bells in a ghost of snow," they remain now only in the memory and words of the poet. Frost compares the dried-up brook to paper on which the poet might compose verse: "Its bed is left a faded paper sheet / Of dead leaves stuck together in the heat— / A brook to none but who remember long" (*CPP&P* 116).

Figure 6.3. Hyla Brook, Robert Frost Farm, July 2023. Photograph by author.

Frost invokes the *Hyla*, a specific species of frog found throughout New Hampshire, here because of its loud and distinctive (and not altogether mellifluous) song.⁹ Rather than a poem of actual human-animal encounter, "Hyla Brook" is a song of absence (or at least seasonal absence). The hylas had once "shouted"; now silent, they have gone "underground" (*CPP&P* 115). The brook, too, has "run out of song and speed" (115). Its song also recalls ancient waters sacred to the poet, like the Pierian Spring, associated with the Muses and poetic inspiration, that would have been meaningful to Frost as an accomplished student of Latin poetry. The poet not only "remember[s] long," but he also fills the void with the contents of his memory: his brook "as it will be seen is other far / Than with brooks taken otherwhere in song" (116). Frost's amphibious song sets itself apart in song from older British and even American models. The aphoristic line that concludes this poem, "We love the things we love for what they are" (116) sounds folksy, but it remains enigmatic. Things are what we make them; their meaning comes from how we recount them, put them into verse, and ultimately value them.

Frost uses another unique species of animal capable of song in "The Oven Bird."¹⁰ Frost's oven bird is "loud," like the hyla, but the speaker tells us that it "knows in singing not to sing" (*CPP&P* 116). Perhaps Frost selected this distinctively North American species because its cry sounds like "teacher, teacher" ("Ovenbird"). Contrary to the speaker's assertion at the beginning of the poem that this bird "is a singer *everyone* has heard" (emphasis mine), the oven bird evidently "is heard more often than the bird is seen" ("Ovenbird"). The male "sings" to attract the female to the nesting territory, and it does so only sporadically during actual courtship. But the female builds a domed nest out of mud, leaves, and animal hairs (resembling an old-fashioned stove) on the forest floor, which is its habitat ("Ovenbird"). Frost's oven bird resembles the *vates*, or poet-priest, who instructs (or "teaches") us something from a place of authority (fig. 6.4).

In writing about this bird, however, Frost distinguishes his poem, with its theme of disillusionment, from the Romantic notion of "transcendent beauty." In *Darwin's Bards: British and American Poetry in the Age of Evolution*, John Holmes argues that Frost's "Oven Bird," which asks "what to make of a diminished thing," and Thomas Hardy's "The Darkling Thrush," which features "an aged thrush, frail, gaunt, and small" (qtd. in Holmes 166), are poems of "disenchantment" (167). According to Holmes, "We only hear

Figure 6.4. Ovenbird (*Seiurus aurocapillus*) bringing nest material to its nest on the forest floor. Nature Picture Library/Alamy Stock Photo.

what the speaker tells us it says," so the poem is "not a portrait of the bird itself at all, but rather of the speaker" (169). Although Frost's poem seems to offer the possibility of "mak[ing] the solid tree trunks sound again," it concludes with a mood of diminishment, whereas Hardy's poem offers a view of nature only "as somehow lessened" (169). By refusing to elevate his midwood bird (another Dantean allusion) to a transcendent height,[11] Frost distinguishes his poem from other writers' bird poems and offers a worldview that is dark and distinctively modern. Frost attributes a kind of wisdom to this bird using anaphora: "He says that leaves are old and that for flowers / Mid-summer is to spring as one to ten"; "He says the early petal-fall is past" (*CPP&P* 116).

Recognizing that springtime is past and summer is here, the oven bird (via the speaker) announces the coming of fall, but not just *any* seasonal fall. It is "that other fall we name the fall," the Fall of Man, linking human and nonhuman animals in a shared biblical mythology. The bird's enigmatic question, "what to make of a diminished thing," is the poet's challenge to the reader and to himself as poet. Our hope is contingent on what we can "make" of our state or what the poet can "make" poetically. The diminishment could also refer to the "fragmentation of habitat as woodlands become developed and large tracts of land are broken up" (Smith 54). After all, the oven bird reminds us that "the highway dust is over all" (*CPP&P* 116), suggesting the intrusion of roads into the forest. Regardless, the poem departs from the use of birds as harbingers of transcendent truth. Read mythopoetically or ecocritically, "The Oven Bird" presents a world in which birds are simply birds and their songs are the words of the poet.

Reading and (Mis)Reading Animals in the Post-Darwinian World

In the previous section, we considered "Hyla Brook" and "The Oven Bird," in which Frost uses animals to express nostalgia and to illustrate the power of poetry. In this final section, we consider three poems—"The Need of Being Versed in Country Things," "A Blue Ribbon at Amesbury," and "Design"—in which Frost reads animals through a post-Darwinian lens without committing himself to a particular scientific or religious stance. Being "versed in country things," for Frost, means understanding country life and the realities of animal husbandry. It also means being well versed, or adept, at fulfilling one's occupation as a farmer or one's office as a poet. In all three of these poems, Frost causes close readers of his poetry to understand that the language humans apply to nonhuman animals usually tells us more about ourselves than it does our fellow creatures. Such is the case with these animals that are "versed," or inscribed into poetry, by a poet who knows the difference between what his speaker, or we, may perceive (or misread) in nature as opposed to what is there.

"The Need of Being Versed in Country Things" appeared in italics as the concluding poem in *New Hampshire* (1923). In it, Frost juxtaposes human loss and nonhuman animal responses to change in the natural world. Such responses come within the context of contemporary debates between science and religion.[12] The poem announces the human tragedy of a farmhouse destroyed by fire, with only its chimney and barn having survived. The poem recounts the passing of time, a lost way of life, and a displaced family. Seeking comfort from the natural world using what John Ruskin famously called the "pathetic fallacy," the speaker receives none. He seeks sympathy from the birds, who have taken up residence in the eaves of the surviving barn and whose "murmur" he initially imagines as being "more like the sigh we sigh / From too much dwelling on what has been" (*CPP&P* 223). Essentially, the speaker seeks an expression of nostalgia from the birds that never comes.

The poem takes an elegiac turn in the penultimate and final stanzas, in which the poet emphasizes the contrasting human and nonhuman animal worlds through repetition: "Yet *for them* the lilac renewed its leaf"; "*For them* there was really nothing sad" (emphases mine). The speaker wants to ascribe human emotions to the birds, as he feels sadness for the loss of the farmhouse and the life it once held, but the birds, "rejoic[ing] in the nest they kept," feel no sadness. The speaker identifies the species of bird in the poem's concluding statement: "One had to be versed in country things / Not to believe the phoebes wept" (*CPP&P* 223). According to the *Audubon Field Guide*, eastern phoebes (or flycatchers), are easily distinguished by their "soft fee-bee song" and are often known to "nest around buildings and bridges" ("Eastern Phoebe"). Frost would have been aware of the name's connection to Phoebus Apollo, Greek god of poetry. Being "versed in country things" also means being attuned to (and writing verse about) the vicissitudes of nature. Kendall reads the poem as "a refusal to mourn[,] which . . . acknowledges even in its denial the occurrences which make mourning inevitable" (335). Ultimately, nature remains indifferent to human and nonhuman animals' suffering, and the poem illustrates the human tendency to anthropomorphize (or misread) nonhuman animals. Refusing to fulfill the task of mourning, the poet gives the final line to the birds.

In "A Blue Ribbon at Amesbury," Frost turns to another type of bird, the white Wyandotte chicken, a breed well known to him (fig. 6.5). Published in *A Further Range* (1936), this poem takes up poultry breeding and competitive showing, activities in which Frost engaged during his Derry years. The poem features a prize chicken, "Such a fine pullet,"[13] that pleases her breeder to no end. The speaker boasts, "She scored an almost perfect bird. / In her we make ourselves acquainted / With one a Sewell might have painted" (*CPP&P* 255). Franklane Sewell (1866–1945) was the foremost illustrator of game birds and fowl of his time, and Frost invokes his name to suggest that this breeder's pullet is *almost* the Platonic form of a chicken. Sewell's colorful and authentic drawings appeared in the *Eastern Poultryman* and *Farm-Poultry*, trade periodicals to which Frost himself contributed articles.[14] Parini records in his biography that the aspiring poet was "extremely fond of his chickens" and "devoted himself to their care" during that time, and he suggests that the events described in this poem recall "the success of one particular bird at a contest he attended" (83).[15]

In this comical poem, Frost contrasts the human world of showing animals at fairs and the animal world of life on the farm, which is this pullet's "home"—a word repeated for emphasis—where she "lingers feeding at the trough." Meanwhile, the breeder of this "fine pullet" has illusions of grandeur (emphasized by Frost through the use of the grand style) in his home, where "He meditates the breeder's art," perhaps suggestive of the speaker in Milton's "Lycidas," who "strictly meditates the thankless Muse."[16] Frost's poultryman goes one step further, however, succumbing to his hubris and fancying the possibility of competing with Mother Eve by "start[ing] . . . a race / That shall all living things displace."[17] The reverie that ensues leads to the poem's conclusion: "The lowly pen is yet a hold / Against the dark and wind and cold / To give a prospect to a plan / And warrant prudence in a man" (*CPP&P* 256). This "lowly pen" is a *made* thing, in Frostian terms, just as a poem is "a momentary stay against confusion" (777). The poem's ending is humble, just as this breeder's and perhaps this poet's ambitions should be in respect to the act of creating.

Frost's most overtly Darwinian statement is his poem "Design," included in *A Further Range* (1936) but written much earlier.[18] This sonnet questions the role of a seemingly benevolent God, who appears amoral—even malevolent—by depicting a spider and moth engaged in a scene of grim ritual.[19] Here, Frost is at his darkest, most "terrifying," if we are to accept the critic Lionel Trilling's infamous assessment of Frost's poetry.[20] In the octave, the speaker discovers a "dimpled spider, fat and white" (*CPP&P* 275), whose innocence is conveyed by its Gerber-Baby-like description and its white color. The flower's name, "heal-all," sounds salutary, but its appearance is less so, as it is spidery like this arachnid. The moth, its rigid prey, is reduced to its "wings" alone, a synecdoche that the poet compares to a "paper kite." This triptych of spider, moth, and heal-all dramatizes nature at work.

The sestet comprises a series of questions that build upon the imagery of the octave through anaphora: "What had that flower to do with being white, / The wayside blue and innocent heal-all"?; "What brought the kindred spider to that height, / Then steered the white moth thither in the night" (*CPP&P* 275)? All of them ponder the designs or plan of God. Even

Figure 6.5. *White Wyandottes to Date*, by Franklane Sewell, 1902. Wikimedia Commons.

Frost's answer takes the form of a question followed by a conditional statement: "What but design of darkness to appall?— / If design govern a thing so small" (275)? Moreover, the poem questions the very question with the use of the conjunction "If." The sestet's leading questions, according to Kendall, "are driven by a crueler irony; they seek to trap the reader as the spider trapped the moth" (360). Such is the "design," or perhaps one of the designs, of this poet for his poem and his readers.

The God of Frost's "Design" is indifferent at best to suffering in the world; at worst, he is malevolent. For Frost's readers, what applies to this moth in the clutches of this spider could apply also to all living creatures, *Homo sapiens* included. In fact, the poem's origins may date to the Frosts' early years in Derry, when they were grieving the recent loss of their son, Elliott, who had died of typhoid fever on July 8, 1900, with both parents at his bedside. Elinor "slipped into a deep depression" after their son's early death, and Robert blamed himself for not having "call[ed] a good doctor sooner" (Parini 68). She responded by saying that "if there was a God, he was malevolent," while he "preferred to think that heaven was simply indifferent to human suffering" (68). Both attitudes toward God (or Nature) can be seen in "Design."[21]

Conclusion

Given the influence of the Derry years on Frost's poetry, let us return to "The Pasture" and the "two Frosts" discussed previously. Much of Frost's verse is set in nature, and his animal poems provide wisdom about our relationships with our fellow creatures and how we see ourselves in them. Much remains ambiguous in these poems, though, and sometimes even ominous. Timothy O'Brien argues that "The Pasture" is "a more intriguing" and "more complex [one] than generally thought" (131), an assessment that makes Frost's decision to make it the prologue to his collections of poetry even more significant. O'Brien suggests that Elinor Frost, who *was* well versed in country things, might *not* have regarded it as so inviting considering its real-world implications (132). After all, Frost's speaker could very well be going out to fetch the calf, separating it from its mother, only to sell it in the market.

In the *Paris Review* interview mentioned earlier, Frost confirmed that some of his poetry was indeed of "a darker mood," as Trilling had stated. "There's plenty to be dark about, you know," Frost stated. "It's full of darkness" (Poirier 111). But his poetry does not have to be *either-or*; it can be *both-and*, as illustrated by the poems discussed in this chapter. According to O'Brien, Frost presents life on the farm as "both idyllic and Darwinian" (133). His poetry can offer a positive "momentary stay against confusion" (whether Darwinian or not), but in its darker iterations it can draw attention to the lack of clarity, or the seeming absence of order, in nature. Perhaps

Figure 6.6. "Robert Frost, poet who is 85 years old today." World-Telegram photo by Walter Albertin. Library of Congress.

this darker view of "The Pasture" would have resonated with Amy Lowell, who admired Frost's "poetic realism" as one of his contributions to modern poetry (81).

Frost sought "to be a poet for all sorts and kinds" (*Selected Letters* 98), thereby fulfilling the role of "everyone's Frost." From a young autodidact of literature and ancient languages to a teacher, a poultry farmer, poet, professor, and lecturer, Frost went on to win four Pulitzer Prizes for his poetry, to be appointed Honorary Consultant in the Humanities at the Library of Congress, to receive a gold medal in recognition of his poetry by Congress, and to deliver a poem at the 1961 inauguration of John F. Kennedy. Despite beginning his poetic career later than most, Frost became the most widely recognized poet in the United States (fig. 6.6).

Frost calls our attention to, but refuses to resolve, either the complexities of human and nonhuman animal behavior or the larger questions implied in his nature poetry. His representation of animals contributes to his appeal as a poet, and his status as a great modern American poet still owes much to his belief that in poetry it is possible to say "one thing" but mean "another." Yet, as he also states, "All metaphor breaks down somewhere. That is the beauty of it" (*CPP&P* 723). In the case of animals, Frost often says one thing in terms of another. Being "versed in country things" himself, Frost versifies "country things" to remind us of the value of work and the work of poetry. Frost's animals call our attention to the ways that we, as readers and human animals, (mis)read both them and ourselves. His birds, horses, chickens, frogs, insects, and all the rest, with their deliberate ambiguity, invite us to regard them—and the nonhuman animals in our lives—with greater attention, sympathy, and perhaps even solidarity.

Notes

I am grateful to John Gruesser, once again, for his advice and expert editorial care. I am also grateful to Mark Jarman and Richard Kopley for their helpful comments on my final draft. I dedicate this chapter with love to my wife, Sharmila J. Patel, for her support and encouragement of my research on Robert Frost.

1. See Virginia F. Smith's helpful "Concordance of Plants" (325–29) and "Concordance of Animals" (331–35) for comprehensive lists of plants and animals—and the poems in which they appear—in *A Scientific Companion to Robert Frost*.

2. Frost's poetry and prose quoted in this chapter, unless noted otherwise, are taken from Robert Frost, *Collected Poems, Prose, and Plays*, edited by Richard Poirier and Mark Richardson, hereafter *CPP&P*, and cited parenthetically by page number(s).

3. Emily Dickinson, "Tell all the truth but tell it slant—(1263)," Poetry Foundation, accessed October 10, 2024, https://www.poetryfoundation.org/poems/56824/tell-all-the-truth-but-tell-it-slant-1263.

4. *New Hampshire* (1923) was the first of four collections by Frost to receive the Pulitzer Prize. The others were *Collected Poems* (1930), *A Further Range* (1936), and *A Witness Tree* (1942).

5. The poem's setting, "Between the woods and frozen lake," recalls the opening of Dante's *Inferno*: "Midway upon the journey of our life / I found myself within a forest dark, / For the straightforward way had been lost" (1.1–23). See Dante, *Inferno*, ed. Matthew Pearl, trans. Henry Wadsworth Longfellow (New York: Modern Library, 2003), 3. Dante's pilgrim allegorically represents humans midway on the journey of "our" life and lost in the dark woods. Frost's little horse seems to remind the speaker, and us, to remain on the right path.

6. See, for example, the Greek myth of Philomela or Frost's myth of the origin of birds' song in "Never Again Would Birds' Song Be the Same" (*CPP&P* 308).

7. Here, one thinks of more famous poems, such as Percy Bysshe Shelley's "To a Skylark," John Keats's "Ode to a Nightingale," or even Edgar Allan Poe's "The Raven,"

from which Frost distances himself (as illustrated in my discussion of "The Oven Bird") even while recalling the tradition of associating birdsong with poetry and poetic inspiration, a topic that would require its own essay to treat adequately.

8. According to the park ranger at the Robert Frost Farm Historic Site, Hyla Brook typically *does* dry up by June. However, on July 7, 2023, as shown in figure 6.3, the brook was in full force, the region having been inundated with rain throughout the previous month.

9. The gray tree frog (*Hyla versicolor*) makes a loud trill and lives near water in forested areas, often under loose bark or in rotting logs during the summer. The hyla hibernates under matted leaves or tree roots in the winter ("Gray Tree Frog").

10. Robert Bly mentions the "nimble oven bird" in his poem "The Slim Fir-Seeds" in *Turkish Pears in August* (2007). Otherwise, the oven bird remains as obscure in verse as it is in nature.

11. Frost's poem can be read as a *recusatio* ("refusal"), a "subgenre of poetic apology" developed by Greek and Roman poets to "express and defend their poetic aims" and to "[redirect] a poetic tradition" (Race 1–3).

12. For treatments of Frost's engagement with science and religion in his poetry, see especially Faggen; Hass; and Pack.

13. The word *pullet* (derived from the Old French *poulet*, diminutive of *poule*, from the feminine of Latin *pullus*), refers to a young hen less than one year old, according to the *New Oxford American Dictionary*.

14. For an introduction to Frost's brief career as a poultry farmer, a collection of his published articles on the subject, and commentary on its possible influence on his writings, see *Robert Frost: Farm Poultryman*.

15. Parini observes that being a poultry farmer also exposed Frost to "the idiomatic speech of local people," which he "tr[ied] on" in the sketches he contributed to these trade periodicals (83) and mastered in such narrative poems as "Home Burial" and "The Death of the Hired Man."

16. See John Milton's "Lycidas": "Alas! what boots it with incessant care / To tend the homely, slighted shepherd's trade, / And strictly meditate the thankless Muse?" Poetry Foundation, accessed October 10, 2024, https://www.poetryfoundation.org/poems/44733/lycidas.

17. This is perhaps another Miltonic reference, this time to the ambition of the fallen angel, Satan, who aspires first to ruin God's creation through the Temptation and Fall and then to repopulate the earth with his own followers. Unlike Milton's Satan, Frost's poultryman is a comical example of everyday hubris. Nonetheless, he can serve as a reminder to the aspiring poet not to aspire to Godhead or, more practically, not to succumb to pride.

18. This poem was first published in 1922 in an anthology of that year's poetry. Only later, in 1936, was it included in *A Further Range*, a delay that Frost attributed to having "forgotten" about it, a contention that most scholars question. The original version, titled "In White," dates to 1912. Some scholars, including Kendall, have attributed Frost's delay to the poem's "theological implications." Others have suggested that Frost wanted to include it with other poems written in a similar vein (Kendall 357).

19. For a similar treatment of an insect's challenge to the religious "reading" of natural events, see Stephen Jay Gould's brilliant essay "Nonmoral Nature."

20. Trilling made this comment at a party celebrating the poet's eighty-fifth birthday. Although Trilling's comments were considered "shocking" by some of his contemporaries, Frost later confirmed the "dark" element in his poetry. Earlier, Randall Jarrell had made a similar assessment of Frost's poetry, regarding this modern quality as one of the poet's many strengths. For an account of the evening, see Parini (408–9).

21. Regardless, Frost pushed through his grief, spurred on by ambition, but he "did not attempt to publish any poetry during his first six years in Derry," devoting himself instead "to the daily rituals of farm life" (Parini 74).

Works Cited

"Eastern Phoebe." *Audubon*, 2024. https://www.audubon.org/field-guide/bird/eastern-phoebe.

Faggen, Robert. *Robert Frost and the Challenge of Darwin*. Ann Arbor: University of Michigan Press, 1997.

Felstiner, John. *Can Poetry Save the Earth? A Field Guide to Nature Poems*. New Haven, CT: Yale University Press, 2010.

Frost, Robert. *Collected Poems, Prose, and Plays*. Edited by Richard Poirier and Mark Richardson. New York: Library of America, 1995.

———. *Robert Frost: Farm Poultryman*. Edited by Edward Connery Lathem and Lawrance Thompson. Hanover, NH: Dartmouth Publications, 1963.

———. *Selected Letters of Robert Frost*. Edited by Lawrance Thompson. New York: Holt, Rinehart and Winston, 1964.

Gould, Stephen Jay. "Nonmoral Nature." *Natural History* 91, no. 2 (1982): 19–26.

"Gray Tree Frog (*Hyla versicolor*)." New Hampshire Fish and Game Department, 2024. https://www.wildlife.nh.gov/wildlife-and-habitat/species-occurring-nh/gray-tree-frog.

Hass, Robert Bernard. *Going by Contraries: Robert Frost's Conflict with Science*. Charlottesville: University of Virginia Press, 2002.

Holmes, John. *Darwin's Bards: British and American Poetry in the Age of Evolution.* Edinburgh: Edinburgh University Press, 2009.

Jarrell, Randall. *Poetry and the Age.* London: Faber and Faber, 1955.

Kendall, Tim. *The Art of Robert Frost.* New Haven, CT: Yale University Press, 2012.

Lowell, Amy. *Tendencies in Modern American Poetry.* New York: Macmillan, 1917.

O'Brien, Timothy. "Taking the Calf to Market in Robert Frost's 'The Pasture.'" *New England Quarterly* 86, no.1 (March 2013): 125–36.

Oerlemans, Onno. *Poetry and Animals: Blurring the Boundaries with the Human.* New York: Columbia University Press, 2018.

"Ovenbird. *Seiurus aurocapilla.*" *Audubon*, 2024. www.audubon.org/field-guide/bird/ovenbird.

Pack, Robert. *Belief and Uncertainty in the Poetry of Robert Frost.* Hanover, NH: Middlebury College Press, 2003.

Parini, Jay. *Robert Frost: A Life.* New York: Henry Holt, 1999.

Poirier, Richard. "Robert Frost: The Art of Poetry, No. 2." *Paris Review* 24 (Summer–Fall 1960): 89–120.

———. *Robert Frost: The Work of Knowing.* New York: Oxford University Press, 1977.

Race, William H. *Classical Genres and English Poetry.* New York: Croom Helm, 1988.

Smith, Virginia F. *A Scientific Companion to Robert Frost.* Clemson, SC: Clemson University Press, 2018.

CHAPTER 7

Marianne Moore's Artist Animals

HEATHER CASS WHITE

The first poems most readers think about when they think about Marianne Moore (1897–1971) are animal poems, partly because there are so many of them. It is not easy to avoid a reference to or detailed description of an animal if one opens Moore's *New Collected Poems* at random.[1] At a conservative count, 41 of the 164 poems it contains are principally about animals, and many more contain references to animals in their texts and titles.[2] More important, however, is the fact that so many of her *best* poems—her most beautiful, most complicated, most philosophically challenging poems—take animals as their subjects. Between 1932 and 1936 Moore published five poems that she affectionately referred to as her "animiles" in a letter to T. S. Eliot.[3] Those poems, "The Jerboa," "The Plumet Basilisk," "The Frigate Pelican," "The Buffalo," and "The Pangolin," are still among the inner circle of those on which her reputation rests. While Moore's attention to animals is too pervasive and multifaceted to be easily summarized, a close look at three of her most celebrated animal poems, "The Jerboa," "The Frigate Pelican," and "The Pangolin," will nonetheless show how her interest in animals was her most flexible tool for articulating her hopes and fears for human beings as a species. Properly understanding those poems, however, requires some historical and personal context around her relationship to animals.

Since her death Moore's place among the essential early modernist American poets T. S. Eliot, Ezra Pound, Wallace Stevens, William Carlos Williams, and Robert Frost has remained secure, and her use of animals as poetic subjects has become one of her three signature contributions to American poetry. The other two are her original stanza forms, in which she measured lines by syllable count rather than traditional poetic feet, and her frequent quotation in the poems of a wide range of prose sources, including advertisements, reference works, and religious tracts. At the same time that it has been celebrated, however, her focus on animals has been the subject of critical anxiety around her poetry's value since she first began publishing. Eliot made it a centerpiece of his introductory essay to her 1935 *Selected Poems*, anticipating the criticism he thought her lengthy poems about animals might provoke. In the wake of World War I and the rise of fascism in Europe, Eliot worried that readers would discount a poet who, for example, devoted thirteen stanzas to a relatively little-known animal like the jerboa. Eliot argued that such doubts would be mistakes, declaring that "the minor subject . . . may be the best release for the major emotions" and that "only the pedantic literalist could consider the subject-matter to be trivial; the triviality is in himself" (xi).

Moore was raised in a household in which animals were regarded fondly and not trivially. One of the first biographical facts one learns about Moore is her family's system of animal nicknames taken from Kenneth Grahame's *The Wind in the Willows*. Moore was called "Rat," or "Ratty," after Grahame's clever, boat-loving Mr. Rat. Her older brother, Warner, the family protector, was called "Badger"; and Mary, their diminutive mother, who raised them alone after the institutionalization of their father, was known as "Mole." Throughout their lives the trio exchanged letters in which they addressed each other this way. In her adult life Moore regarded animals with a trained scientist's eye, after majoring in biology at Bryn Mawr (having failed to make the grades she needed to be an English major). Her training in biology gave her a fine hand for sketching, and beautiful drawings of animals, plants, and other subjects enliven her notebooks (fig. 7.1).

Moore, who lived in Brooklyn in close quarters with her mother until the latter's death at age eighty-five, had a mixed history with pets. An eye-opening pair of anecdotes involving cats suggests the peculiar mixture of tenderness and pragmatism Moore was capable of where domestic animals were concerned. She tells the first anecdote herself, in a 1910 letter to her mother and brother written while she was away at a summer job in the Adirondacks. She was working at Melville Dewey's Lake Placid Club, a retreat offering "'comfort, rest, and attractive recreations' to the 'overworkt and nervously exhausted'" (Leavell 103). A stray cat was caught on the property. Hearing about the camp director's wish that it be disposed of, Moore asked if she could have it. Upon getting the cat to her room, she chloroformed it herself

Figure 7.1. *Tritonis charonia*, drawn by Moore at Norfolk, Virginia, August 28, 1936.

and then "tied the pussy to the table spreading the paws, made an anterior and a posterior transverse cut and a longitudinal, dissected one leg thoroughly." Her motivation in the affair was curiosity about the leg of a lame dog her family once owned; she reported satisfaction with the project in which "I saw what I wanted to know" (*Selected Letters* 84).

Linda Leavell, in her biography of Moore, *Holding On Upside Down*, tells a more complicated cat story. In the early 1920s a neighbor gave Moore and her mother a stray kitten. They named him "Buffy" and wrote doting letters about his beauty and cleverness to Moore's brother. After keeping him for a month, however, Moore wrote to Warner: "Mole got chloroform and a little box and prepared everything and did it while I was at the library Monday, and nothing could have been more exact. . . . But it's a knife in my heart, he was so affecting and scrupulous in his little scratchings and his attentions to our requirements of him. . . . A seemingly comfortable life in a shut up room would not be good for any cat so we were kind, but having had him so long made the deed seem foul" (Leavell 207). Even four years later, Leavell reports, Marianne could not go near the pier where they buried Buffy.

Home not being an option, Moore relied instead for her contact with animals on her regular attendance at zoos and circuses, her reading in scientific books and journals, and above all her reading in the *London Illustrated News*. Founded in 1842 and published weekly until 1971, the *London Illustrated News* was a newspaper dedicated to articles that featured more pictures than text. It covered current events, the arts, popular and natural science, and other subjects of general interest.[4] Moore was an avid reader for many years, finding inspiration in its pages for images and whole poems. Her notebooks are filled with information she learned there and with sketches of its photographs. Photography and writing were Moore's pathway to animals, a pathway that may be felt in the way her poems regard animals with fascination and reverence, but also from an unmistakable distance, both emotional and physical. Few poets have been more alive to the multiple meanings animals have had for human culture throughout its history, but there are no Moore poems (with the possible exception of "Peter," about a friend's cat) about the particular human-animal intimacy that enriches many people's everyday lives and is celebrated in, for example, Mary Oliver's poems about dogs. The story of Buffy suggests Moore was capable of such feeling, but she did not write from it in any explicit way. Notably, the "animiles," her five major animal poems, are about animals Moore never saw in person.

Nevertheless, Moore's poems about animals embody most of what is distinctive about her worldview. They communicate her beliefs about beauty, and right and wrong, and the purposes of art; they teach by example the principles by which she lived. The first such principle, foundational for all the others, is her commitment to accurate observation. Any given description of an animal in a Moore poem will be composed of precise sensory details that are fresh and surprising because of her genius for the unexpectedly accurate comparison. Elephant skin is "fibered over like the shell / of the coconut" (41). A cat asleep is "flat- / tened out by gravity, as if it were a piece of seaweed tamed / and weakened by / exposure to the sun" (45). The plumed basilisk, a lizard that runs on water, "when captured [is] stiff / and somewhat heavy, like fresh putty on the hand" (109). A monarch butterfly has "sunburnt zebra-skin" and "round unglazed china eyes" (138). An ostrich revolves his "comic duckling head" with "compass- / needle nervousness" (151). A skunk wears "black and white chipmunk / regalia" (163). The kiwi has a "rain-shawl / of haired feathers" (169). Such examples are essentially endless.

Moore's arresting, evocative descriptions are part of the aesthetic pleasure of her work but are also one of its primary ethical coordinates. Moore believed in knowing what was to be known about the world that surrounds us and in interacting with others on the basis of education and the enlarged self it grants. She was not at all concerned about sounding lofty. Her most famous poem, called simply "Poetry," begins with a notorious dismissal. "I too, dislike it: there are things that are important beyond all this fiddle." By "all this fiddle" she means poetry, the art to which she was dedicating her life. Having established herself on that ground, however, she makes an immediate concession: "Reading it, however, with a perfect contempt for it, one discovers that there is in / it after all, a place for the genuine" (27). According to Moore, finding the "genuine" in poetry requires that poets be fundamentally unfinicky, alive to the artistic potential in everything they encounter: animals such as bats, horses, and elephants, yes, but also intransigent critics, baseball fans, statisticians, business documents, and schoolbooks. "All these phenomena are important" (27), she emphasizes—it is up to the poet to see them that way and make the reader see it too. Moore consistently represents that capacity to see as a matter of training rather than giftedness. We don't need to be special, she suggests, but we do need to work if we want to act as moral agents in the world. In her strongest statement on the subject of right education, her 1932 poem "The Student," she advocates for fearlessness in the pursuit of knowledge, since "one would not care to hold opinions that fright // could dislocate" (96). As in "Poetry," she asks us to widen our perspective on where knowledge may be found. "The football huddle in the vacant lot," she reminds the reader, "is impersonating calculus and physics and military / books; and is gathering the data for genetics" (95–96).

When "The Student" presents examples of particular things we ought to learn, however, it turns, characteristically, to animals. We should know "the difference between cow / and zebu; lion, tiger; barred and brown / owls; horned owls have one ear that opens up and one that opens down. // . . . The penguin wing is ancient, not degenerate" (96). Moore's observation about the penguin wing, like many other such images, passes quickly in the poem where it appears but has deep roots in her philosophy (fig. 7.2). Examined closely, it shows the connections between her interest in animals and her belief in learning as an ethical act. Although it is not set in quotation marks, the phrase quotes a 1931 article from the *London Illustrated News* on which she took notes. By calling the penguin wing "ancient" rather than "degenerate," the article corrects a popular misperception of penguins as badly adapted birds who have lost their (presumed) former ability to fly. Learning the history of the species shows that they have instead been adapted since the beginning for flight in water instead of air. Only our ignorant expectations of birds make them seem "degenerate" to us.[5] Education, Moore emphasizes, restores our appreciation for their dignity and skill.

MARIANNE MOORE'S ARTIST ANIMALS | 141

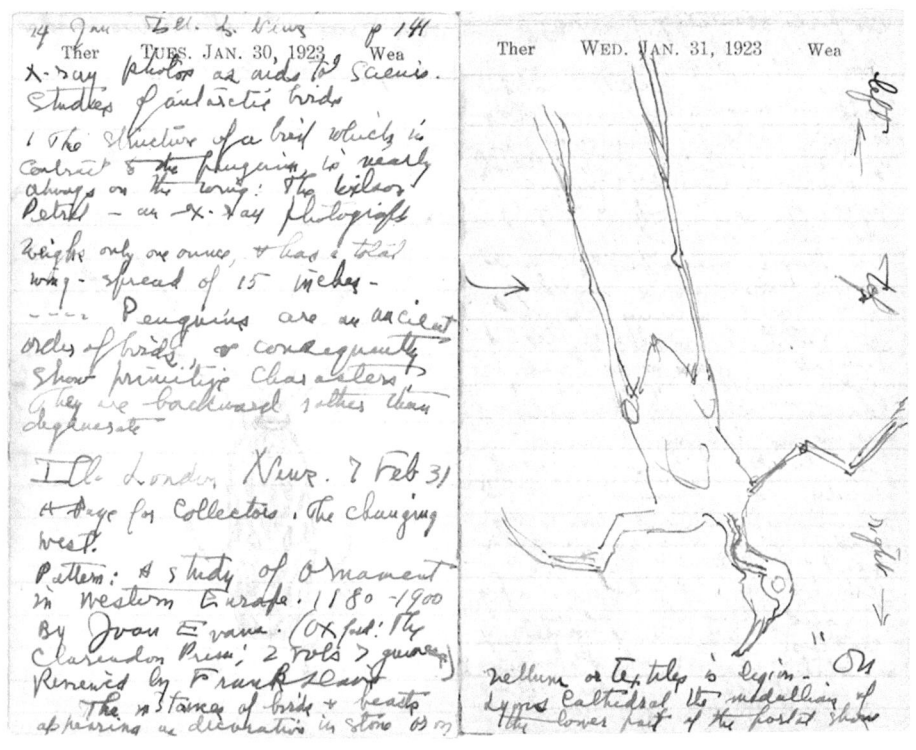

Figure 7.2. Page from Moore's notebook with notes on an article about arctic seabirds in the *London Illustrated News*, January 24, 1931.

Moore's poem "The Jerboa" shows clearly Moore's belief in what animals have to teach us if we are willing to observe them closely on their own terms. At 162 lines it is one of her longest poems. It is also one of her most formally intricate, composed of twenty-seven stanzas of six lines each, rhymed in an AABCDD pattern, each with the same pattern of syllable count by line: 5–5–5–11–10–7. The poem is divided into two long sections, titled "Too Much" and "Abundance," which juxtapose two ways of living in the same desert. "Too Much" details the decadence of ancient Egyptian royal life, while "Abundance" celebrates the simplicity of the jerboa's neighboring, innocuous life in the sand dunes. "Too Much" takes particular exception to the Egyptians' easy sense of ownership over animals that ought to be free: "they looked on as theirs, / impalas and onagers, // the wild ostrich herd" (100). It notes with distaste that their harnessing of animal power, using "dapple dog- / cats to course antelopes, dikdik, and ibex" (100), is of a piece with the structures by which history has come to know them: these were a people who "understood / making colossi and / how to use slaves" (100). When the Egyptians are not enslaving animals, they are turning them into grotesque toys, "put[ting] goose-grease / paint in round bone boxes . . . incised with the duck wing . . . ; ke[eping] in a buck / or rhinoceros horn, / the ground horn; and locust oil in stone locusts" (101). In all, the poem identifies Egyptian animal husbandry as the sign of a larger, ongoing human entitlement that seems "right to those with, everywhere, // power over the poor" (101).

The jerboa, by contrast, has power over nothing. This "not famous" "small desert rat" lives with "no water, no palm-trees, no ivory bed" (102). It is so fleet of foot, making "fern-seed / foot-prints with kangaroo speed" that one can hardly see it at all as it goes about its nocturnal gleaning (104). Unlike the Egyptians, it leaves no trace more lasting than those footprints in the sand and amasses nothing but the food it needs to survive. Yet, as the poem phrases it in one of Moore's customary double negatives, "one would not be he who has nothing but plenty." Unlike the Egyptians who have "too much" of everything, the jerboa lives in the abundance of his "happiness," which is the sand itself (fig. 7.3). "O rest and / joy, the boundless sand" the poet exclaims, merging briefly with the attitude of the animal who "honors the sand by assuming its color" (104). She in turn honors the jerboa's resourcefulness and perfect adaptation to its environment by paying minute attention to its physical details: the nap of its fur "directed / neatly back and blending / with the ear," the "fine hairs on the tail, / repeating the other pale / markings," the remarkable power of its "match-thin hind legs" from which it "launch[es] / as if on wings" (103). For Moore the jerboa's sublime freedom is an explicit rebuke to human beings, who, she contends, trap ourselves in gaudy prisons of our own making. We have colossi; the jerboa has happiness. We restlessly alter our environments to suit our perceived needs; the jerboa honors his environment and is in turn strengthened by it, "silvered to steel by the force / of the large desert moon" (103). She directs the jerboa's moral lesson specifically at the practice of enslavement in the first stanza of "Abundance" by pointing out the legacy of imperialism in the Egyptian continent's name. "Africanus," she notes tartly, "meant / the conqueror sent / from Rome. It should mean the / untouched" (102). Like the sand-brown jumping rat she admires, she suggests, African people were, are, and should always have remained "free-born" (102).

Figure 7.3. *Jaculus orientalis*, the greater Egyptian jerboa. Photo by Nikolay Usik. Wikimedia Commons (CC BY-SA 3.0).

As "The Jerboa" shows, it is easy to like the animals Moore describes, and even easier to generally approve of the principles they embody for her. Most readers, however, usually want art to ask more from them than easy approval. A lasting criticism of Moore's poetry is that it pays only selective attention to animals, cherry-picking their more attractive qualities for praise. The poet and critic Randall Jarrell (1914–65), who was otherwise Moore's ardent fan and remains one of her best readers, offered a rare rebuke on the point:

> The way of the little jerboa on the sands—at once true, beautiful, and good—she understands; but the little shrew or weasel, that kills, if it can, two or three dozen animals in a night? the little larvae feeding on the still living caterpillar their mother has paralyzed for them? Nature, in Miss Moore's poll of it, is overwhelmingly in favor of morality, but the results were implicit in the sampling—like the *Literary Digest*, she sent postcards only to the nicer animals. (179)

Jarrell's rebuke stings because it is true, at least for the most part. There are minor exceptions: Moore likes the fierceness of cats, noting with pleasure that "an animal with claws wants to have to use / them" (46), and also likes the mercilessness of the mother bird confronted by a cat, waging "deadly combat" on and half-killing it with "bayonet beak and / cruel wings" (137). Nevertheless, Jarrell is so right in general about Moore's bias in choosing animals as moral examples for human beings that the one instance when she didn't display that bias is all the more interesting. Most of Moore's animal poems involve models for ethical human behavior. When she wanted to consider the ethical dimensions of a human *artist's* behavior, however, she looked to an animal without the virtues she usually admired.

Her 1934 poem "The Frigate Pelican," published two years after "The Jerboa," celebrates a bird it frankly calls "dishonest." Moore had in mind John James Audubon's portrait of the frigate pelican, also known as the magnificent frigatebird (fig. 7.4). Moore describes the portrait's particulars exactly in imagining the bird's flight: "the toe / with slight web . . . and very long wings / . . . feel / the changing V-shaped scissor swallow- / tail direct the rigid keel" (112). The frigatebird's "dishonesty" starts with his abundance of aliases. The poet remarks that although he is commonly called the frigate pelican, "he is not a pelican." His other noms de guerre include "hell-diver, frigate-bird, and hurricane-bird"; the mere sight of him flying close to the waves is a "storm omen" (112). He lives up to his tempestuous reputation in his feeding behavior. Although his agility means he "should be seen / fishing," it is not easy to catch him at it. Instead "he appears to prefer // to take, on the wing, from industrious cruder-winged species / the fish they have caught, and is seldom successless" (112). This thieving bird, who literally takes the food from the mouths of others who work harder for it, makes no apologies for his nature or his lack of domestic virtues. He cannot be bothered with the mundane: "Make hay; / keep the shop . . . were a less / limber animal's mottoes" (113). Nor does he have anything in common with nurturing fairy-tale animals. Unlike "the stalwart swan that can ferry the / woodcutter's two children home," "this one . . . would not know Gretel from Hansel," let alone feed its children from blood torn from its own breast, as the true pelican does in early modern European myth.

Soaring overhead, the frigate pelican keeps "at a height / so great the feathers look black and the beak does not / show" (113). Moore notes without rancor that "it is not retreat but exclusion from / which he looks down"

Figure 7.4. Robert Havell Jr., John James Audubon, Frigate Pelican, from the book *Birds of America*, 1835, hand-colored engraving and aquatint on paper, sheet: 38 x 25⅜ in. (96.6 x 64.3 cm), Smithsonian American Art Museum, Museum purchase through the Robert Tyler Davis Memorial Fund, 1980.91.

(113); her admiration for the lawless and asocial bird centers entirely on the "the height and . . . the majestic / display of his art" (113). Unlike the jerboa, who teaches a right way of living, the frigate pelican is Moore's model of the artist, a necessarily solitary figure who stands apart from social norms, claiming the freedom he needs to exercise his skill. She makes the point explicit by comparing the bird to "Handel— / meant for a lawyer and a masculine German domestic / career" who instead secretly studied the harpsichord, "and never was known to have fallen in love" (113).[6] Aloof from a sense of obligation to others, and hidden by the altitude he can achieve, the frigate pelican has enviable skill and unique power, "able to foil the tired / moment of danger, that lays on the heart and lungs the / weight of the python that crushes to powder" (114).

Enchanted as she is by the bird's flight, calling it a "reticent lugubrious ragged immense minuet" (114), the poet observing him is herself earthbound, watching the display rather than joining it. She feels a conflict he doesn't, which she expresses late in the poem as a mistranslation followed by a question: "*Festina lente*. Be gay civilly. How so?" (114). *Festina lente* is an adage of military origin in use since ancient Roman times. It is usually translated "make haste slowly," a paradox that advises cautious speed, careful advancement that does not make progress that will easily be lost. Moore translates it differently, with the competing agendas of art making and citizenship in mind. John Slatin puts it succinctly: "The question for Moore is how to make poetry (which the Provençal poets called *gai saber*, 'the gay science') a civil art—how to write freely without sacrificing her sense of community" (205). The frigate pelican may be able to sustain itself and execute its aerial flourishes on a diet of stolen fish, but Moore, who in "The Jerboa" disdainfully sums up the imperial logic of the ancient Egyptians in six words, "the bee's food is your food" (101), will not.

Nor will she give up, however, on a love for the dazzling products of single-minded labor. The jerboa's triumph is how little he takes from his environment and how perfectly he fits himself to it. Moore admires those qualities in a member of a living community, but if everyone had them, we would have no breathtaking flights of daring to thrill us or examples of extreme beauty and skill to aspire to. Even in "The Jerboa" Moore's fascination with the products of artistic ingenuity shows through in flashes. The Egyptians made colossi of which she disapproves, but they also made expertly crafted materials like "yarns dyed with indigo," "red cotton," and "flax which they spun // into fine linen / cordage for yachtsmen" (101). Those morally off-putting duck-shaped boxes and stone locusts themselves have their charm for Moore, one of whose early poems is about just such an artifact: "An Egyptian Pulled Glass Bottle in the Shape of a Fish" (1924). Anything well made was attractive to Moore, who was an indiscriminate admirer of good craftsmanship. Her friend and fellow poet Elizabeth Bishop liked it in her: "Marianne was intensely interested in the techniques of things—how camellias are grown, how the quartz prisms work in crystal clocks; how the pangolin can close up his ear, nose, and eye apertures, . . . how the best pitchers throw a baseball, how to make a figurehead for her nephew's sailboat. The exact way in which anything was done, or made, or functioned, was poetry to her" (135).

Bishop's third example, of the pangolin's defensive bodily structures, refers to Moore's poem "The Pangolin." There is no more important poem in the Moore canon for investigating the complexity of her relationship to animals and their meaning for her as an artist. One immediate sign of the importance of "The Pangolin" is that she first published it in a book rather than a periodical and that she titled the book for it: *The Pangolin and Other Verse* (1936). She included "The Pangolin," with only light revisions, in all subsequent editions of her collected and complete poems, and it has become a touchstone poem for readers of Moore. The pangolin itself, a nocturnal, intricately armored anteater native to Asia and sub-Saharan Africa, may be the closest thing we have to Moore's mental self-portrait, an image of the removed but not antisocial "artist-engineer" she aspired to be (fig. 7.5). Moore found in the pangolin's uniqueness an image for her own. She was a born poet who "disliked" poetry, who advocated both blending in like the jerboa and standing out like the frigate pelican, who despised the thoughtless exercise of power over animals but was drawn to the products of the civilization that conspicuously engaged in it. These contradictions left her with

Figure 7.5. *Manis temminckii*, ground pangolin. Courtesy of David Broussard. Wikimedia Commons (CC BY-SA 2.0).

difficult questions: Should she aspire to dazzle or to deflect attention from herself as her poems' maker? When are the tools of poetic form itself—its meter, its rhyme—mere "fiddle," and when are they necessary parts of the "majestic display" of one's art? How, in a phrase, can the poet "be gay civilly"?

Each Moore poem is a new answer, but "The Pangolin" is a particularly rewarding one. The structure of the poem is simple: the first six and a half stanzas describe the pangolin as it makes its nighttime feeding rounds, and the final three and a half describe man. The poet compares the two animals, each with characteristic strengths and weaknesses, and concludes that man is an endearing, if somewhat self-defeating, mammal worth our continued attention. As in "The Jerboa" and "The Frigate Pelican," Moore focuses in "The Pangolin" on what a creature's works tell us about its values. Moore wrote about animals as though "spirit creates form" (37), treating their own bodies as their most telling creations. She often treats the animal world as a system of moral emblems, in which what an animal looks like tells us its meaning for human life. "The unity of / life and death," for example, "has been expressed by the circumference / described by [the elephant's] / trunk" (41–42), while the "thin glass shell" encasing the eggs of the paper nautilus is "her perishable / souvenir of hope" (158).

Few other animal forms evoke as much affectionate wonder from Moore as the pangolin's; her descriptions of its remarkable shape are both precise and funnier than usual. Midway through the poem Moore reprimands "simpletons" who "thought [the pangolin] a living fable," but there is something fantastic to her as well about the mysterious animal, moving silently through the moonlight "peculiarly, that the out- / side edges of his / hands may bear the weight and save the claws / for digging" (141). Looking at him, Moore often does not see an animal at all, describing his armor as "scale / lapping scale with spruce-cone regu- / larity" and twice calling him an "artichoke." Seeing him "serpentined about / the tree," she thinks of "the fragile grace of the Thomas- / of-Leighton-Buzzard Westminster / Abbey wrought iron vine"; watching him walk, she notes his "not un-chainlike, machine- / like form and frictionless creep" (141, 142). Moore calls the pangolin "the night miniature artist- / engineer," wondering if he is "Leonardo's indubitable son?"[7] Her question suggests that she sees the pangolin as an artist whose creation is himself. Taking ants as his raw materials, he transforms them into his own ingenious form, notable for its power to delight

MARIANNE MOORE'S ARTIST ANIMALS | 149

Figure 7.6. A pangolin rolled into a defensive ball. Photo by Stephen C. Dickson. Wikimedia Commons (CC BY-SA 4.0).

the eye. Like Leonardo, however, the pangolin is also an engineer for whom beauty is rooted in perfect functioning (fig. 7.6). The pangolin may have the "fragile grace" of wrought iron, but he also has the strength of iron and the "frictionlessness" of well-machined parts. He is able to "draw / away from / danger unpugnaciously" because the natural world has little to threaten him when he "rolls himself into a ball that has / power to defy all effort / to unroll it" (141–42).[8]

The poem makes the transition between its focus on the pangolin and its focus on man by means of a quiet pun. The Asian genus of pangolin is called *Manis*, a word that visually resolves easily into "man is." Moore takes advantage of this coincidence, noting that when it stands on its hind legs, the "manis [has] certain postures of a / man" (143). We share other features too, such as the "warm blood, no gills, / . . . and a few hairs" that make us both mammals (144). Unlike the unique, frictionless pangolin, however, in Moore's view we humans are at best clumsy imitators of the animals around us. "A paper-maker / like the wasp; a tractor of food-stuffs, / like the ant;

spidering a length / of web from bluffs," man, she writes, "slaving / away to make his life more sweet, leaves half the / flowers worth having" (143). By contrast the pangolin's discreet nighttime artistry is perfect, because in his unassuming way he is the perfect master of his instrument: he repels swarming ants with "leg and body plates" that "quiver violently" and is helped by his giant tail, a "graceful tool, as prop or hand / or broom or axe, tipped like the elephant's trunk with special skin" (142). Man, meanwhile, can hardly get anything right. We declare ourselves "writing- / master to this world" and then "write errror with four / r's" (143–44). We claim to be "not afraid of anything" and yet go "cowering forth, tread paced / to meet an obstacle / at every step" (144). Lacking the pangolin's armor and tail, we do the best we can, "serge-clad, strong-shod." Like lesser diurnal pangolins we too endure "exhausting solitary / trips through unfamiliar ground" (141). What the pangolin does "impressively" (141) by moonlight we do awkwardly by the light of day. Always "the prey of fear," always "thwarted by the dusk, work partly done," the poem concludes, we persist by looking outward for strength, praising the sun that "anew each day" "comes into and steadies [our] soul" (144).

Moore first published "The Pangolin" in 1936, in a limited-run fine press edition. By the time she included it in her next book with a major publisher, *What Are Years* (Macmillan, 1941), the world had changed, as had Moore's life. She was deeply troubled by the approach and advent of World War II and had been corresponding since the late 1930s with a British friend, the writer and philanthropist Winifred ("Bryher") Ellerman, about attacks on European Jews. She was also in correspondence with Ezra Pound, whose poetry she valued and whose fascism and anti-Semitism she deplored. The prospect of US involvement in the conflict was personally fraught for Moore, whose navy chaplain brother was likely to be deployed. She was also facing worries in her professional and home life. In 1940 Macmillan, her longtime publisher, declined a novel she had been secretly working on for more than a decade, and she was having trouble publishing her new poems, which were rejected by the *New Yorker* and the *Atlantic*. Meanwhile, her *Selected Poems*, the book on which T. S. Eliot had sought to cement her international reputation, was remaindered. Perhaps worst of all, her mother's health was increasingly bad, and much of Moore's time was taken up by caring for her. (Mary Moore eventually died in 1947.)

After the 1940s Moore was a different poet from the one she had been in her youth. She sought publication in new venues with more general readerships, she redoubled her commitment to engaging with the pressing moral issues of her day as she saw them, and she worked, insofar as a poet of her innate complexity was able, to simplify her poems' methods and presentations. In other words, if the guiding question of her earlier work had been how to "be gay civilly," her later work suggests that when the two values conflicted, civility won. Most famously Moore registered her changing poetic priorities by extensively revising her earlier poems. It had always been her habit to make changes to her poems right up until the moment of publication, and often to make changes between published versions, as when she included a poem in a book that had first appeared in a periodical. The scale of the alterations she was making by the late 1940s was new, however. She was working on her *Collected Poems* (1951), a book that, despite its title, omitted ten poems she had previously published in books and cut six of her best-known poems by one-third to one-half.

Along with these attention-getting changes to her work, there is a subtle shift in her use of animal subjects. From the 1940s onward there are fewer poems devoted primarily to animal subjects, and those that exist are less likely to center on wild animals than on those cultivated for human use, such as reindeer introduced to Lapland, arctic oxen grown for qiviut, and racehorses. The "free-born" wildness of animals like the jerboa, the frigate pelican, and the pangolin no longer drew her. Instead, she admired the usefulness of animals performing well the functions humans have assigned to them. The most striking evidence of this change in her perspective is the contrast between two of her poems about elephants. She published the first, "Black Earth," in 1918. The second, "Elephants," appeared in 1943. "Black Earth" is spoken in the first person by an elephant contemplating the spiritual meaning of his physical body. The elephant is interested in his form not as anyone else might see it but as it feels from the inside. His tone is authoritative: "Openly, yes, // I do these / things which I do, which please / no one but myself" (41), and frank about his experience of strength: "Black / but beautiful, my back / is full of the history of power" (41). He is untroubled by the marks of time, by his tough skin "cut / into checkers by rut / upon rut of unpreventable experience" (41) or "the sediment of the river which / encrusts my joints" (41). The elephant believes that "the / patina of circumstance can but enrich what was / there to begin // with" (41) and that what was there to begin with in him is a searching after "spiritual poise" (42), an inner peace rooted in grounds other than mere "pride" (42). Part of his confidence is knowing that he doesn't know what those grounds are, just as he doesn't know the answers to large questions: "what is powerful and what is not?" (41). Instead of looking away from such issues, he remains on alert, "ears sensitized to more than the sound of / the wind" (42).

The elephant subject of the 1943 "Elephants" sees the world differently. Unlike his relative in "Black Earth," he lives in captivity, serving as an elaborately dressed participant in ceremonial processions to the "Temple of the Tooth" in Kandy, Sri Lanka. Entrusted with the relic of Lord Buddha's tooth itself, the elephant has absorbed the Buddha's lessons on patience and nonattachment better than his human rider. Although infinitely more powerful physically than the man he serves, the elephant is content to "expound the brotherhood / of creatures to man the encroacher" (165). Protected by the elephant's presence, its "prone mahout" "sleeps as soundly as if . . . // invisibly tusked, made safe by magic hairs!" (164). The elephant's choice to submit to its human rider signals a larger shift in emphasis with respect to the character qualities Moore herself admired. In her earlier animal poems, she responds to independence of mind and self-determination, as in the elephant of "Black Earth." By 1943 the elephant of "Elephants" makes a whole philosophy out of being "a life-prisoner but reconciled," who once "resisted, but is the child // of reason now" (164). While Moore was always fond of self-effacing animals, like the jerboa blending in with the sand, as well as the more defiant ones like the frigate pelican, the element of captivity and the connection she saw in it to "reason" were new.

As a younger poet Moore was sensitive to the charge of "unreasonableness" and its synonyms: "eccentricity," "curiosity," "idiosyncrasy." As a poet who sounded like no one else, who followed no precedents in her meter and wrote about subject matter not commonly associated with poetry at all, she was used to hearing such terms in descriptions of her own work. She felt their trivialization keenly and confronted it directly from two flanks, sometimes disputing the terms and sometimes owning them. In yet a third, very early elephant poem, "Diligence Is to Magic as Progress Is to Flight" (1915), she repudiates the charge by analogy, praising elephants' thick skin for its ability to "endure blows," which in turn proves the seemingly ungainly animals, rather like her own seemingly ungainly stanza forms, to be "prosaic necessities, not curios" (18). In "Black Earth" she again concentrates on the elephant's protective skin as an image of her own resistance to anything or anyone who would try to standardize her art. At the end of the poem, however, she embraces the judgment she repudiated in "Diligence Is to Magic." "Will thick skin be thick," she asks rhetorically, "to one who can see no beautiful element of unreason under it?" (42). Moore's history of identifying with elephants, and in particular with their freedom to answer to no one else's sense of beauty or purpose, suggests how multilayered a statement she is making when she describes the harnessed, captive elephant of 1943 as "the child // of reason now" (164). Her earlier elephants stood for self-determined strength, and a poet who insisted that what looked strange or ephemeral to others was the permanent source of her poetry's value. Her later elephant decides instead that true wisdom is submission to necessity, even when the necessity is to serve beings weaker than itself.

Moore herself was aware of the changes her poetry underwent as she got older and the philosophical shift that underlay them. Nor did it escape her that her animal poems were the most sensitive register of those changes. In 1951 she wrote a family member that the only poems of hers she still liked were a handful written in the late 1940s, dismissing the rest as "my 'cats and dogs' of former days" (Leavell 344). Moore's readers have tended to disagree strongly with this assessment, liking those "cats and dogs" very much, each in its way as irreplaceable and unique as the animal that inspired it. A philosophy of resigned acceptance and selfless service may have been valuable to Moore in her later life. She certainly wrote many fine poems, including "Elephants," with it in mind. Without the earlier, less reasonable, wilder poems that preceded them, however, they might not have been preserved at all. Moore survived and may even have come to love her domestication, but the place for the genuine she made in American poetry has the freedom of her "animiles" at its heart.

Notes

1. This is the volume to which I will refer by page number throughout. It contains the most accurate and complete record of the poetry Moore published during her life at a national and international scale and reprints her poems in their earliest versions, which are not easily found elsewhere. The story of Moore's revisions to her own poems over time is fascinating in its own right. Readers who are curious about it can read the *New Collected*'s "Introduction" and essay on "Editing the Poems."

2. The proportion of her work dedicated to animals rises dramatically if one takes

into account her 1954 translation of *The Fables of La Fontaine*. I have limited my discussion here to her original poems, but it is worth bearing in mind that when she wasn't composing her own, she turned to intense engagement with another writer's animal poems. At its present length this chapter cannot begin to do justice to the range of excellent criticism on Moore's use of animals. My own understanding has been informed by, in addition to the critics I do cite, the foundational work of Bonnie Costello, Celeste Goodridge, Jeanne Heuving, Cynthia Hogue, Cristanne Miller, and Laurence Stapleton, as well as more recent work by Luke Carson, Christina M. Colvin, Fiona Green, Elizabeth Gregory, James Dennis Hoff, Jennifer Leader, Cliff Mak, and Sabina Sielke.

3. Moore to T. S. Eliot, July 2, 1934, Rosenbach Museum and Library, Philadelphia. In fact, there was a sixth "animile," called "Pigeons." Moore published it in *Poetry* magazine in 1935 but never included it in any subsequent collection. She left no record of why she made that decision, and it was unfortunate, as "Pigeons" is the equal of her other long animal poems of the mid-1930s. It is included in the *New Collected*.

4. Victoria Bazin's study *Marianne Moore and the Cultures of Modernity* explores Moore's use of this periodical in depth and connects her responsiveness to its primarily visual emphasis to a wider cultural moment in which "imaginary acquisition and collection" had become "a legitimate and valued social activity for women" (14).

5. The word "degenerate" links this scientific clarification to a larger national anxiety around the possibility of the "American race," meaning Americans of Anglo descent, losing fitness for survival by overconsuming European art and culture. Robin Schulze's study *The Degenerate Muse: American Nature, Modernist Poetry, and the Problem of Cultural Hygiene* details the impact of this anxiety, beginning in America in the late nineteenth century, on Moore and other poets.

6. In *The Web of Friendship*, Robin Schulze makes the persuasive case that Moore had Wallace Stevens in mind while she wrote the poem, although Stevens himself, a lifelong insurance man, husband, and father, was un-frigate-bird-like in his nonwriting life.

7. In later reprintings of the poem she removes the doubt from the comparison, replacing the question mark after "son" with a period (see White 118).

8. The exception to the natural world's powerlessness in the face of his armor is man. The World Wildlife Fund ranks the pangolin as the world's most trafficked mammal and has put all eight species on its critically endangered list. There is good reason to think that the next generation of Moore's readers will see "The Pangolin" as an epitaph to an animal having become a fable like the dodo.

Works Cited

Bazin, Victoria. *Marianne Moore and the Cultures of Modernity*. Burlington, VT: Ashgate, 2010.

Bishop, Elizabeth. *Prose*. Edited by Lloyd Schwartz. New York: Farrar, Straus and Giroux, 2011.

Carson, Luke. "Republicanism and Leisure in Marianne Moore's Depression." *MLQ: Modern Language Quarterly* 63, no. 3 (2002): 315–42.

Colvin, Christina M. "Composite Creatures: Marianne Moore's Zoo-Logic." *ISLE: Interdisciplinary Studies in Literature and Environment* 24, no. 4 (2017): 707–26.

Costello, Bonnie. *Marianne Moore: Imaginary Possessions*. Cambridge, MA: Harvard University Press, 1981.

Goodridge, Celeste. *Hints and Disguises: Marianne Moore and Her Contemporaries*. Iowa City: University of Iowa Press, 1989.

Green, Fiona. "'The Magnitude of Their Root Systems': 'An Octopus' and National Character." In *Critics and Poets on Marianne Moore: "A Right Good Salvo of Barks,"* edited by Linda Leavell, Cristanne Miller, and Robin G. Schulze, 137–49. Lewisburg, PA: Bucknell University Press, 2005.

Gregory, Elizabeth. "Marianne Moore's 'Blue Bug': A Dialogic Ode on Celebrity, Race, Gender, and Age." *Modernism/Modernity* 22, no. 4 (2015): 759–86.

Heuving, Jeanne. *Omissions Are Not Accidents: Gender in the Art of Marianne Moore*. Detroit: Wayne State University Press, 1992.

Hoff, James Dennis. "Marianne Moore, John Dewey, and the Aesthetics of Animal Life." *Texas Studies in Literature and Language* 61, no. 3 (2019): 311–33.

Hogue, Cynthia. *Scheming Women: Poetry, Privilege, and the Politics of Subjectivity*. Albany: State University of New York Press, 1995.

Jarrell, Randall. *Poetry and the Age*. New York: Knopf, 1953.

Leader, Jennifer. *Knowing, Seeing, Being: Jonathan Edwards, Emily Dickinson, Marianne Moore, and the American Typological Tradition*. Amherst: University of Massachusetts Press, 2016.

Leavell, Linda. *Holding on Upside Down: The Life and Work of Marianne Moore*. New York: Farrar, Straus and Giroux, 2013.

Mak, Cliff. "On Falling Fastidiously: Marianne Moore's Slapstick Animals." *ELH* 83, no. 3 (2016): 873–98.

Miller, Cristanne. *Marianne Moore: Questions of Authority*. Cambridge, MA: Harvard University Press, 1995.

Moore, Marianne. *New Collected Poems of Marianne Moore*. Edited by Heather Cass White. New York: Farrar, Straus and Giroux, 2017.

———. *The Pangolin and Other Verse*. London: Curwen, 1936.

———. *What Are Years*. New York: Macmillan, 1941

———. *Selected Letters*. Edited by Bonnie Costello. New York: Penguin, 1997.

———. *Selected Poems*. New York: Macmillan, 1935.

———. *What Are Years*. New York: Macmillan, 1941.

Schulze, Robin G. *The Degenerate Muse: American Nature, Modernist Poetry, and the Problem of Cultural Hygiene*. New York: Oxford University Press, 2013.

———. *The Web of Friendship: Marianne Moore and Wallace Stevens*. Ann Arbor: University of Michigan Press, 1995.

Sielke, Sabine. "Strange Animals in Stylish Habitats: Marianne Moore's Poetry Revisited." In *An Eclectic Bestiary: Encounters in a More-than-Human World*, edited by Birgit Spengler and Babette B. Tischleder, 123–40. *Human-Animal Studies* 20. Bielefeld, Germany: Transcript: Verlag für Kommunikation, Kultur und soziale Praxis, 2019.

Slatin, John M. *The Savage's Romance: The Poetry of Marianne Moore*. University Park: Pennsylvania State University Press, 1986.

Stapleton, Laurence. *Marianne Moore: The Poet's Advance*. Princeton, NJ: Princeton University Press, 1978.

White, Heather Cass, ed. *Adversity & Grace: Marianne Moore, 1936–1941*. Victoria, BC: ELS Editions, 2012.

CHAPTER 8

Noticing Other Species—and Our Own—in Elizabeth Bishop

CALISTA MCRAE

Years ago, at the Houghton Library, I opened Elizabeth Bishop's copy of W. H. Auden's *Thank You, Fog* (1974) to this three-line poem:

Many creatures make nice noises,
but none, it seems,
are moved by music. (23)

In one of the very few margins where she leaves something more than a question mark or exclamation point, Bishop has penciled: "*No—seals are—*."[1] She would know; in "At the Fishhouses" (1947) she remembers Nova Scotia evenings where she watched a seal who "was interested in music" (*Complete Poems* 65; hereafter *CP*). Since he was evidently "a believer in total immersion," she "used to sing him Baptist hymns." The seal would maintain his position in the water, watching her. Then, after "disappear[ing]," he would resurface "with a sort of shrug / as if it were against his better judgment" (65). That last phrase epitomizes the cautious interplay of perception and imagination in Bishop's depictions of animals. Her interpretation of the shrug is self-deprecating—was there something questionable about her singing, or her theology, or simply her attempt at interaction?—but she is also serious in her interest about what might have been in the seal's mind. She is alert to her projecting, to how she is drawn to construe the "shrug"; this recursive self-questioning is central to her work as a whole, particularly her poems about the nonhuman world.

Before seeing Bishop's marginalia in that volume of Auden, I took her assertion that the seal was "interested in music" as more playful than truthful. But about forty years after she wrote in that book, researchers documented a captive California sea lion learning to bob her head to songs, quickly responding to changes in tempo (Cook et al. 2013). Previously, rhythmic synchronizing was thought to be something only vocal mimics could do; the sea lion showed that less vocally varied animals can also be literally "moved by music." Bishop seems to have met another pinniped who responded to human music. Her emphatic underlining—"seals are"—suggests she felt strongly about Auden's slightly complacent generalization (the kind of distinction that humans are apt to make quite readily) and felt impelled to correct it.

Bishop is a famously good observer, and her four collections of poetry are full of poems including animals that a number of critics have recently discussed. Most have focused on the innovation and care with which she represents the captivating, opaque lives of nonhumans. Priscilla Paton describes Bishop's "critical anthropomorphism" (211), a combination of empiricism and identification that seeks "to approach, though rarely to know completely, what behaviourists could not: the mindedness of an animal" (211). Johanna Hoorenman examines how Bishop "puts nonhuman animal experiences on par with human experiences of the world, rendering fluid any boundaries between human and animal emotions" (481), and Michael Malay examines Moore's and Bishop's "'shared predicament' of writing about animals in ways that did not detract from or compromise the creatures' otherness" (38). Marianne MacRae, similarly, charts how Bishop "maintains [animals'] absolute otherness" while also "creat[ing] a sense of recognition between human and animal" (369). Andre Furlani characterizes "Bishop's representation of the animal" as "an art of translation—fidelity and acknowledgement more than identification and strict correspondence" (439). Most recently, Toshiaki Komura has argued, "At the core of Bishop's animal poetry is a sense of loss—whether it is the loss of connection with nature, the sublimity of natural landscape, the immediacy of human-animal relations, or the ability to understand animals" (6). Each of these readers speaks to Bishop's interest in depicting animals as truthfully as possible, even while going beyond what can be objectively observed.

I am interested in a related facet of Bishop's ethical attention to animals—the tensions in her work that arise from *noticing*. If you are a good observer, you are likely to observe things you wish you didn't, such as scenes of unwitting or outright cruelty: as Bishop wrote Anne Stevenson in 1964, "Lack of observation seems to me one of the cardinal sins, responsible for so much cruelty, ugliness, dullness, bad manners—and general unhappiness, too" (*Prose* 413). And you may also observe that even your own moments of perceptiveness are interspersed with moments of obliviousness: for every animal that delights Bishop, there is an animal affected by human impercipience. After an overview of Bishop's general, lifelong interest in animals, I begin with two unfinished, unpolished drafts that not only show especially starkly her concern for other species but also her struggle to render this concern in a way that satisfied her aesthetic criteria: these drafts show the difficulties of writing about animal welfare, or animal cruelty. The rest of the chapter focuses on her published poems, where she attends both to the animals closest to humans—those most obviously and constantly subject to our actions—and to putatively "wild" animals, who also end up affected by us.

I move from passages where Bishop looks mainly at the behavior of her fellow humans to those where a figure standing for Bishop herself is implicated: Bishop persistently notices her own potential for imperceptiveness, and indeed her own complicity in animal harm. While she does write about her literal, direct roles in other animals' lives, I am especially interested in the moments when she seems troubled by the ethics of writing poetry about animals. On the one hand, she knows that animal poetry makes no difference to an actual suffering creature, and, on the other hand, she knows that animal poetry stems from a difficult human impulse, the impulse to make contact with another creature even to its potential detriment.

Bishop's Creatures

Often what Bishop notices is an animal's presence itself. She records beings generally missed and what they are doing. In "At the Fishhouses," for instance, she describes the seal's behavior, not his looks—she does not mention whether he is dark or light, spotted or blotchy or silvery, sleek or scruffy. And she doesn't specify species; most of her animals stay at family or genus (a seal, a dog, birds, sparrows, a sandpiper).[2] Rather, instead of focusing on an animal's physical features with relentless, sustained closeness, she observes through multiple senses and on multiple scales. She takes in what is nearby, like a pig's "light-lashed" eyes amid the "breathing and thick hair" (71), and what is distant, like circling buzzards seen as "stirred-up flakes of sediment / sinking through water" (32).[3] In all but a few instances, the length of her look is brief—perhaps in part because encounters with real animals, or at least with wild animals, themselves tend to be brief. Bishop's animals are almost always seen directly, whether they are bobbing in the ocean nearby, or glimpsed through the window of a bus, or a barely perceptible movement in the sky. As Malay writes, these are "personal encounters with animals, occasions recounted in insistently experiential and individual terms" (81–82).

Bishop lived with various animals over her life. Some of the first were three tiny bantam chickens given to her by her paternal grandparents. The bantams brought her happiness in an otherwise lonely time: "When one hen pecked at some cornmeal on the enamel table, and made miniature but hen-like sounds, I could have cried with pleasure" (*Prose* 97). Her grandparents also had Beppo, a Boston terrier who (they did not realize at the time) exacerbated her asthma and eczema. Later, Bishop never kept a dog herself and avoided staying with people who had dogs, but she kept many cats and birds. In the 1930s she had a cat named Minnow; in 1966, she wrote that her cat Tobias "still flourishes, age 15," with a new Siamese companion (*One Art* 448); she owned a copy of Margaret Gay's *How to Live with a Cat* (1946). She bought three owls in Rome in her late twenties; they did not get past customs in the United States. Her letters from the 1950s are full of anecdotes about Sammy the toucan, a gift from a neighbor in Brazil. Near the end of the 1960s, missing the birds of Brazil, she kept a myna, likely a common hill myna.

Perhaps because she spent so much time in the company of other species, Bishop was alert to the ways humans affect them. The animals in her poems are defined by her persisting concerns about human-nonhuman relations: almost every one of them is in some way marked by humans. Although she came of age before the start of the modern environmental and animal rights movements (in the late 1960s and early 1970s), she imagined the specter of extinction early: her story "The Last Animal" (1934) is set in a world so nearly devoid of nonhuman life that a child cannot identify the one mammal found, confined, and exhibited (*Prose* 475–79). Occasionally, she avoided meat: "(I was also a vegetarian until after college, I think!—and I revert to it every once in a while. I don't advocate it or even believe in it—but they drive the cattle to market here, and after each encounter with one of the cattle trains . . . I give up meat again for a week or so.)" (*Prose* 432). That explicitly intermittent practice, over decades, suggests an awareness both of animal suffering and of the insignificance of a personal gesture. And it suggests self-consciousness about the strong feelings that often accompany sympathy for animals: the "or so" reads like an attempt to make the sentence casual, to avoid any hint of self-righteousness.[4]

In her poems, Bishop's troubled sense of human-caused damage is at its most explicit in two unfinished drafts. The first begins by watching freighters, and then whalers, pass the coast of Rio:

> *They kill the whales with cannon. Can't*
> *they leave the blessed whale in peace?*
>
> . . .
>
> *The holds are full of dying animals.*
> *Don't worry, we'll consume them all*
> *sooner or later. But how you have*
> *roiled up the water! Splash . . .*
> *What was that sound? The trash.*
> *They throw it overboard at night. (Edgar Allan Poe and the Juke-*
> *Box 131, hereafter EAPJB)*

The anger, sarcasm, and revulsion audible in the lines quoted here show Bishop as appalled both by whaling and by her own inadvertent involvement; the draft's pronouns move from "they" to "we," then to "you" and back to "they." She spans multiple scales of damage: the suffering of the increasingly far from "blessed" whales,[5] but also the general environmental destruction, exemplified by the sneaky disposal of waste.

In a 1978 interview with Elizabeth Spires, when asked about unfinished work, Bishop remembered this poem: "It's about whales and it was written a long time ago, too. I'm afraid I'll never publish it because it looks as if I were just trying to be up-to-date now that whales are a 'cause'" (*EAPJB* 328–29). When she wrote her draft in the early 1960s, few people thought much about whales. In 1964, the English journalist E. S. Turner could describe the harpooning process ("blow a hole in a man, anchor a rope in his wound and make him pull a loaded barrow for an hour or two—that is a rough comparison" [14]) and then conclude, "Not many of us care about whales; they are majestic and mysterious, but readily overlooked. It is easier to boycott a circus than to boycott soap" (15). Bishop, actually seeing the whaling vessels pass, was not able to overlook them, though she admitted she was part of the "we" benefiting from their deaths. That dual noticing—of others' cruelty and of one's own, more passive role—will run through several of the poems discussed later in this chapter.

Bishop's remarks in the Spires interview also suggest a reluctance to seem part of a movement or, more generally, to write poems that may seem to be trying to make a change in the larger world. In another interview, she and George Starbuck discuss the word "tract"; at one point Bishop says, "Tract poetry . . ." and trails off, as if skeptical about whether poems that seek to make an unambivalent, unambiguous point are actually any good. Poems about animal harm tend to have a degree of definiteness that Bishop would have seen as lacking the complexity associated with lyric. Bishop was unwilling to publish the whale poem in its current form: it makes a point about animal exploitation so immediately that it feels more certain and closed than Bishop's poems usually do.

A similar problem—and an example of Bishop's sense of her own culpability—comes through in an unfinished elegy for Sammy the toucan. Sammy was likely a keel-billed toucan (fig. 8.1), since Bishop wrote in a letter, "Most of him is black, except the base of the enormous bill is green and yellow and he has a bright gold bib" (*One Art* 234). He died because of an insecticide Bishop had been told was safe for birds. In the elegy, Bishop addresses the dead toucan, telling him, "I was afraid your life was boring"; the draft ends, "I loved you, and I caged you" (*EAPJB* 179). The aggressive parallel of that last sentence makes explicit Bishop's knowledge that human affection can be selfish.[6] As she did with the whale poem, it's possible to see why Bishop abandoned the draft, though she tried repeatedly to finish it. The tidy decisiveness of that final sentence does not capture Sammy's comedy or resilience and unpredictability, or anything else about him described in the poem's earlier lines: self-condemnation overwhelms those qualities and other aspects of their years together.[7] The poems discussed in the pages that follow, in contrast, find ways of allowing for Bishop's full range of perceptions, in all their conviction and confusion.

The Nonhumans near Humans

Some of the most intense—and happiest—human-animal interactions in Bishop's poems involve dogs. As animals that have adapted relatively well to domestication, Bishop's dogs are often described with phrases that call attention to their more or less comfortable positions vis-à-vis humans. Dogs

Figure 8.1. Keel-billed toucan (*Ramphastos sulfuratus*) in Costa Rica. Photo by Ttschleuder. Wikimedia Commons (CC BY-SA 3.0).

often seem at ease in human groups. Several dogs appear in "The Moose," for example, starting with a collie "supervis[ing]" one family headed off to the bus (*CP* 170). That slightly comic verb, with its meanings of watching and of actually being in charge of something, suggests how a collie can look—and very literally feel—protective of its human pack. Later the speaker thinks of a quiet house where everyone has settled in for the night, including "the dog / tucked in her shawl" (172). That image is of an animal who is at home, whom someone has helped make snug—in "her" shawl, the one she is used to, the one that has come to be hers. The word "shawl" is unlike the more expected "blanket" in that shawls are not simply pieces of cloth; they are articles of clothing, worn by women. The dogs of "The Moose," sketched in just a few words, seem to be members of human families. So does the "dirty dog, quite comfy" on a sofa, near equally greasy men, in "Filling Station" (127).

But domestication is not always thoroughly untroubled. There is a paragraph in Bishop's memoir "The Country Mouse" about how Beppo, the terrier from her childhood, once punished himself for having vomited in the hallway: he had gone to sit in a closet where people had put him for other offenses, though never for vomiting (*Prose* 91). In acquiring the human sense of guilt, he has become what Priscilla Paton calls "a slightly absurd domesticated creature" (200). While Bishop kept plenty of pets, she shares with other midcentury poets some unease about how humans may change a species over generations, and a single dog over the course of his lifetime. She seems to have believed, or suspected, that some dogs' now-almost-innate desire to please humans has left them susceptible to human traits.

The presence or absence of animal guilt—or rather of what Bishop calls "shame"—is central to "Five Flights Up," the final poem of Bishop's final book, *Geography III* (1976). In each stanza, either a "little dog" or a "little bird" or both appears, going about their lives as day begins (*CP* 181). In the third of the four stanzas, the dog "runs in his yard," *his* yard, another possessive adjective (like "her shawl") that gives the dog a place where it is at home. And then Bishop hears "his owner" scolding him for reasons unknown: "You ought to be ashamed!" (181). But the dog, in response or lack of response, "bounces cheerfully up and down; / he rushes in circles in the fallen leaves" (181). Watching from above, Bishop's speaker remarks, "Obviously, he has no sense of shame," and then, in the last, more wistful lines, finally introduces the poem's only "I":

> *He and the bird know everything is answered,*
> *all taken care of,*
> *no need to ask again.*
> *—Yesterday brought to today so lightly!*
> *(A yesterday I find almost impossible to lift.)* (181)

What is immediately notable is the dog's imperviousness to those things that can weigh on the minds of human beings. The dog reacts to the scolding by "bounc[ing]" and running "in circles," around and around—full of energy, but not moving forward in time. He shrugs time off, while the human "I," faced with the prospect of knowing what has happened and is still happening, is saddled with an almost physically heavy "yesterday."

That contrast risks idealizing nonhuman life, by reducing it to enviably unconscious freedom. But as the marginalia in Auden suggests, Bishop would want to avoid such a reduction. She does so by what she includes in the first two stanzas. At the beginning of the poem, the dog "barks in his sleep / inquiringly, just once" (181). While he seems joyfully on autopilot in the third stanza, his "bark[ing] in his sleep" introduces more uncertainty about what a dog might experience, despite his lack of "shame." Dreams intrigued Bishop: the one little bark here suggests that the dog, too, is capable of dreaming and perhaps dreams to cope with anxiety or another emotion.

Then there is the little bird, also asleep in the first stanza. As is the case elsewhere in Bishop, this bird seems to be a figure for creatures beyond the direct reach of humans. Bishop first wonders if he, too, might dream, like the dog: but in stanza 2, still sitting on "his usual branch," he "seems to yawn." A yawn is both a physical reflex and a sign of an affective state. At least in humans, it suggests a less-than-engaged state of mind and thereby *a* state of mind: to be bored or sleepy is to imply an inner life that is not fully stimulated. By including the dog's hypothetical dream and the bird's apparent yawn, Bishop avoids turning all of nonhuman life into symbols of freedom from self-consciousness, or into Cartesian beast-machines, or into human doubles.[8] Cautious about all such generalizations, she gestures to several possible inner states for other species. And at the same time, she recognizes the ways one particular animal (the one who says, "You ought to be ashamed!") is driven to mold others.

Discrepancies in how we treat animals are especially visible across different species, as "Under the Window: Ouro Prêto" suggests. Under the window lies a centuries-old water source that "all" "agree" is especially cold. "All," it turns out, encompasses not just people but other passersby: "Donkeys agree, and dogs, and the neat little / bottle-green swallows dare to dip and taste" (153). The word "all," normally reserved for humans, here broadens to literal animal enjoyment: many creatures gravitate toward this water. And yet there is still a separation between species considered nearer to and farther from us: while thirsty donkeys and dogs and a truck crew "agree," the swallows remain otherworldly darters, not part of the general agreement that earthbound creatures share. While this species dividing line departs from the predictable human-nonhuman one, it is related to it; the speaker's perspective is inevitably rooted in the human.

As part of the poem's larger recording of what is heard and seen from the window, later tercets take in other glimpses of animals, mainly those used in some way by humans:

> *Six donkeys come behind their "godmother"*
> *—the one who wears a fringe of orange wool*
> *with wooly balls above her eyes, and bells.*
> *They veer toward the water as a matter*
> *of course, until the drover's mare trots up,*
> *her whiplash-blinded eye on the off side. (154)*

These donkeys—more specific than the donkeys seen as mammals amid other mammals near the poem's beginning—are marshaled into rows. Their lives are largely defined by how humans want them to work, and even to look: the "godmother" is ornamented with bright, somewhat cluttered decorations ("wooly" and "wool," "balls" and "bells"). Those emblems of fairly innocuous human intervention are juxtaposed with the permanent alteration made to another working animal, the horse who has been partly blinded by incidental brutality. The faint sense of constraint, epitomized by the fringe and lash, is quietly reinforced by the fact that these animals are not actually shown drinking but only as heading or trying to head toward the water. While Bishop does not make animal abuse anything like a focal point in "Under the Window," she does remark on it by moving from an image for our shared animal nature to an image of how we control other animals.

Of the several snippets of conversation overheard in "Under the Window," the last suggests both the common ground shared by humans and nonhumans and the distinctions humans make. Someone below says, "For lunch we took advantage / of the poor duck the dog decapitated" (154). Here Bishop pushes together a fleeting moment of nominal pity ("poor duck") and the clinical word "decapitated," in a comically grotesque sentence.[9] But the most incongruous phrase is "took advantage of," a roundabout and revealing way of saying "we ate": Bishop has in mind the advantages humans have and take—despite how close they are to other animals—and the kinds of contradictions that Turner pointed out in his 1964 book: "In our attitude to animals we are hopelessly, perversely inconsistent. There have been fox-hunters who revolted at the idea of performing animals. . . . All of us applaud the trouble taken to tranquillise and lift hippos from the sites of dams, but none of us care whether rats are killed humanely or cruelly" (12). While Bishop was nowhere near as outspoken, she perceived the irrational selectiveness of human empathy that Turner catalogs.

Animals that are literally more distant from humans nevertheless continue to be altered by them to varying degrees. In a letter to Robert Lowell, Bishop mentions planning a trip to "sanctuaries where there are auks and the only puffins left on the continent, or so they tell us" (*Words in Air* 6; hereafter *WIA*). This trip, or one like it, makes its way into the opening of "Cape Breton," where

> *the razorbill auks and the silly-looking puffins all stand*
> *with their backs to the mainland*
> *in solemn, uneven lines along the cliff's brown grass-frayed edge,*
> *while the few sheep pastured there go "Baaa, baaa."*
> *(Sometimes, frightened by aeroplanes, they stampede*
> *and fall over into the sea or onto the rocks.) (CP 67)*

Initially, this scene is slightly comic, thanks both to the pose of the birds and the children's-book transcription of "Baaa" (fig. 8.2). But then there is the parenthetical stampede, with its ending "onto the rocks" (67); though the speaker has gone to see the uncommon seabirds on their isolated island, she also sees the grisly deaths of sheep. Claire Seiler explains that a space near this island was used for bomb practice: military planes were what frightened the sheep (76). But in Bishop's poem, no reason is given for the planes. The complete lack of explanation makes the deaths seem even more unnecessary. It suggests that while the islands have been turned into sanctuaries for rare birds, other animals are still vulnerable to human thoughtlessness. That Bishop puts the fact into parentheses, technically making it inessential, underscores human obliviousness. In fact, it might suggest the speaker's own thoughtlessness—that she seems to mention these casualties as an afterthought.

The Appeal of Animals, and Their Appearances in Poems

What the parentheses do in miniature—implicate the speaker herself, suggest her interest centers on the sight of the rare auks and puffins rather than on the panic of the sheep—is writ large in several of Bishop's most famous poems. "The Armadillo," where animals fleeing a frivolous, human-made fire are illuminated and described in terms of what has destroyed their home, captures not only the aesthetic draw of wild animals but our slowness to recognize harm. As Christopher Spaide writes, the poem "announces the

Figure 8.2. Atlantic puffins. Image by Joanne Goldby. Wikimedia Commons (CC BY-SA 2.0).

shortcomings of a worldview that elevates aesthetic remove over empathic connection" (60). Its frightened animals are relentlessly aestheticized, as are the "frail, illegal" fire balloons themselves, whose "paper chambers flush and fill with light / that comes and goes, like hearts" (*CP* 103); beautiful, delicate phrases permeate both the injured and the injuring.

Bishop ends with an italicized stanza that calls into question such aesthetic reactions. In the words of Heather Bozant Witcher, this stanza involves "at least a dawning awareness on the part of the poetic speaker of the impact of humanity on the natural world" (281). As part of this belated awareness, the stanza may rebuke the "dreamlike mimicry" (*CP* 104) of the poem itself, how the poem has treated everything as beautiful. And Bishop may even suggest that this poem has exploited and disguised the animal suffering it depicts.[10] Although she had doubts about "tract poetry," here she may admit the limitations of its opposite.

Wariness about poetry might also be detected in two other of Bishop's most famous poems. One—tonally quite different from the prettiness of

"The Armadillo"—is "Pink Dog." Its subject, a street dog in Rio, is pink because she is hairless, something the speaker recoils at: "Oh, never have I seen a dog so bare! / Naked and pink, without a single hair" (190). She addresses the dog directly, in loudly rhyming tercets. Knowing the dog does not have "rabies" despite the fears of other passersby, she continues: "You are not mad; you have a case of scabies / but look intelligent. Where are your babies?" (190). Most saliently, the poem supports an allegorical reading about all the cruelty human beings are capable of inflicting on people who do not fit in. The dog, Bishop's speaker explains, is likely to end up like the "beggars" who are thrown into the "ebbing sewage" as part of the city government's attempt to clean up.

But Bishop's choice of constant direct address calls attention to the nominally concerned, exclamatory voice itself—a voice that keeps talking to and asking questions of a creature who cannot answer. Bishop may suspect that, in her concern for animals, she resembles a particular kind of white woman present in English literature at least as far back as the risibly sentimental prioress in Chaucer's *Canterbury Tales*, who weeps for suffering dogs and mice. While her racial awareness as suggested across her poems is conflicted and blinkered, in this poem she seems to feel a discomfiting link between poetry, concern for animals, and white femininity.[11] It is not just that Bishop sees herself as the type of woman who has the leisure to worry about animals: she has the leisure to write poems about animals. She is working in a genre of privilege, with a voice that keeps "just talking" (as her speaker says of the concerned citizens of Rio).

This uneasiness about poetry—perhaps especially about "animal poetry," in a world hostile to animals—pervades "The Fish," another of Bishop's most anthologized poems. Her speaker opens with a surprisingly stock claim about the size of the fish caught, thus positioning the poem in a much broader history of fishing discourse. Bishop soon describes at length everything that is happening to or on the fish's tattered integument: his skin "hung in strips / like ancient wallpaper" (42). Regardless of the species—there has been some debate,[12] and Bishop herself suggested that it was a fish now called the goliath grouper (fig. 8.3)—most saltwater fish do not lose their scales in strips as this one has, and peeling is generally taken for granted as being a bad sign.[13] One bulletin from the US Fish Commission, describing a disease outbreak in salmon in the late 1870s, refers to how the diseased skin "assumes the consistency of wet paper and can be detached in flakes, like a slough" (Walpole and Huxley 430). At a hearing a hundred years later, a fish expert referred to peeling as a symptom that would obviously keep people from unintentionally ingesting the *Pfiesteria* organisms then feared to be circulating: "these toxins are so lethal to fish so quickly, that they cause fish to look bad, to become diseased, and the skin peeling, and so forth" (*Oversight Hearing on* Pfiesteria 19). If peeling seems to be a marker of illness, Bishop's fish is compromised, especially given the lice, weeds, and barnacles also on his skin, and his lack of resistance: "He didn't fight. / He hadn't fought at all" (*CP* 42).

Whatever has happened to this fish's scales, the very act of grasping him for longer than needed is likely to exacerbate the damage. In Hemingway's "Big Two-Hearted River," the protagonist dreads the way ignorant fly fishermen touch trout with their dry hands, leaving dead fish in their wake.[14] It's not clear how this fish is being held, but in general, the less handling the better with fish, and regardless of the hold, what follows in the poem is protracted. Anat Pick has observed, "The time that Bishop's detailed descriptions require (in the real time of fishing and reading) makes the poem a ticking clock" (419): the fish is out of water for a palpably and potentially fatally long time.

Near the beginning of this chapter I asserted that most of Bishop's animals are not described at great length. "The Fish" is an exception. Its metaphors are as extensive as the barnacles, lice, and weed obscuring his actual body. And the details are both an aesthetic pleasure—like so much in "The Armadillo"—and a quiet commentary on the underlying harm that has made the poem possible. Over the next fifty lines and approximately two minutes of reading time, the fish is an occasion for strikingly inventive description. The speaker's aesthetic experience is made grammatically central, while the experience of the fish is tucked into dependent clauses and phrases: "While his gills were breathing in / the terrible oxygen" is subordinated to "I thought . . . ," not letting one stop to appreciate the word "terrible" and its etymological implications of causing terror. Later sentences parallel "I thought" with "I looked," "I admired," and finally "I stared and stared," so that the grammar is grounded in the speaker's slow, careful acts of observation and association.

The metaphor-heavy surface of "The Fish" has several effects. First, it charts the potentially deadly amount of time that is passing. Second, it demonstrates how the speaker transforms—or rather, obscures—the marks

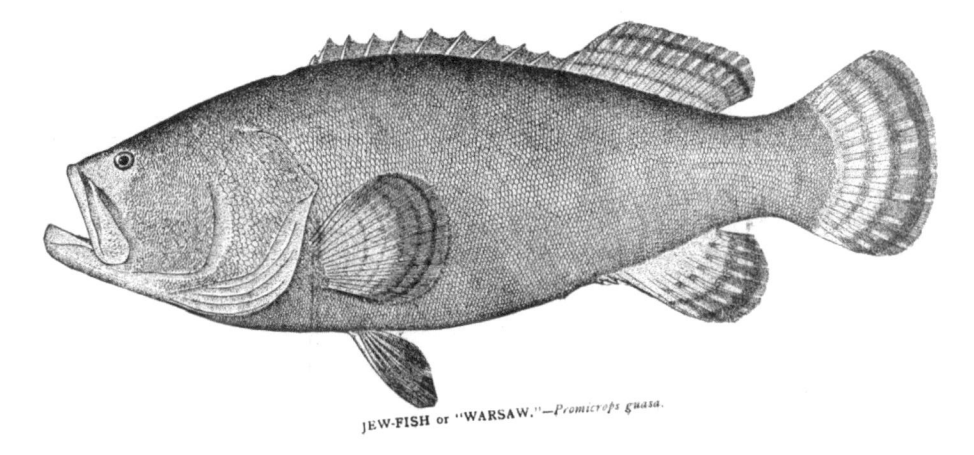

NOTICING OTHER SPECIES IN ELIZABETH BISHOP | 171

Figure 8.3. The Atlantic goliath grouper. From *W. A. Perry's American Game Fishes*, 1892. Wikimedia Commons.

of suffering, especially human-inflicted suffering, turning them into something gentler and prettier. Bishop's speaker sums up the hooks in the fish's mouth with the following images:

> *Like medals with their ribbons*
> *frayed and wavering,*
> *a five-haired beard of wisdom*
> *trailing from his aching jaw. (CP 43)*

First the hooks turn into medals, making the fish an old, venerable soldier who has grown wiser through what he has survived. And then, as if not satisfied with that metaphor, Bishop renders the hooks as something more natural—a beard, something that comes through time, not through anything she or other humans do. These three lines of comparison get priority over the "aching jaw," even though the sentence ends with and thereby emphasizes that reference to long-standing, dull, accumulating pain. This recognition of suffering is surrounded by a self-conscious attempt to make the hooks seem something other—something less disturbing to the beholder—than what they are.

After recognizing the aching jaw, Bishop's speaker tells us, "I stared and stared." In the lines that follow, however, everything except the fish is stared at, as the speaker's gaze moves to the materials of the "little rented boat."

Bethany Hicok has noted that "the rainbows caused by the oil from the boat" direct "our attention to human intervention in the natural world" (62). The five pieces of equipment Bishop's speaker looks at—engine, bailer, thwarts, oarlocks, gunnels—parallel the five hooks the fish is already stuck with. And the polluting oil is made beautiful, just as the grim, weaponlike metal is turned into an affirmation of the fish's toughness. David Kalstone has written that "The Fish" is "filled with the strain of seeing—not just the unrelenting pressure of making similes to 'capture' the fish, but the fact that the similes themselves involve flawed instruments of vision, stained wallpaper, scratched isinglass, tarnished tinfoil" (87). That attention to "seeing" could be extended to imperceptiveness: to seeing brilliantly, but not fully understanding what one sees, and not fully understanding the implications—for the fish—of one's painstaking observations.

The poem does not end in an explicit revelation about the harm Bishop's speaker may have contributed to or an enigmatic reprimand like that of "The Armadillo," though it evokes the intense aesthetic reaction of "The Armadillo" in its exclamation of "rainbow, rainbow, rainbow." Instead, the last line—"and I let the fish go"—is so quick as to raise questions about what happens. Coming after extended, close attention to the fish, there is nothing about how he was extricated from the new, sixth hook, the one mentioned at the poem's opening, where it was "fast in a corner of his mouth." But even in this minimizing, Bishop does comment on her role as a person who "captures" the fish in two senses. More broadly, she documents a moment of intense interest—one of the increasingly rare interactions that humans have with other animals—and its inadvertent consequences. She captures a layer of imperceptiveness amid intense, close observing; how we unwittingly or wittingly harm other creatures, sometimes because of our admiration; and how poetry itself may exploit other species.

Conclusion: What Are You Projecting?

Such moments of observing—observing actual human interventions into the nonhuman world, including one's own but also one's poetic tendencies—occur throughout Bishop's work. She is unobtrusively preoccupied with human misreading and projection. I want to end by glancing at "Roosters," a long poem about birds who can be vicious to each other (fig. 8.4). Here Bishop must write about an animal that is territorial in the way many animals are, including many human ones, but whose aggression is exacerbated by people. As Thomas Travisano writes, "These militaristic roosters have been trained to fight and even kill one another by their cock-fighting masters" (76); Dave Schaefer's history of Key West notes that Key West roosters "are descendants of roosters brought by Cuban immigrants for cock fighting, which went on into the 1950s" (55). Roosters in general have long been used in human allegories—as symbols of bravery, braggadocio, virility, even Christianity—something this poem certainly continues.[15]

For the poem's speaker, the roosters' form of masculinity, how they loudly and violently assert their territory, is as obnoxious as most others. But we should notice the human interference—Bishop's speaker's interference—too. The speaker's attitude comes through strongly when she describes "their cruel feet" and how they "glare // with stupid eyes" (*CP* 35). Technically, neither of the adjectives is accurate. Feet can't be cruel and eyes can't be stupid.[16]

Figure 8.4. *Key West Rooster on the Boardwalk*. Photo by Carl Vizzone. Flickr (CC BY-NC 2.0).

These words emphasize not just hostility but the way a human's feelings toward an animal can infiltrate even the descriptions of physical features. The poem is full of tiny verbal intrusions: the roosters' feathers become, in the speaker's frustrated eyes, a "vulgar beauty."[17] And when the speaker hears their "scream[s]" and asks them, "Roosters, what are you projecting?" (36), the question stems both from frustration and a more-probing desire to understand the roosters' inner lives. What might go through their heads? And how could a human ever begin to guess, given how much we have projected onto them?

Near the very end of "Roosters," Bishop includes a transient image of an animal not—or not yet—affected by humans: "the tiny / floating" swallow, whose "belly" is caught and "gild[ed]" by the sunrise. This swallow, like the swallows of "Under the Window: Ouro Prêto," is remote from people. One of the most permanently airborne of birds, here it seems to "float" effortlessly, without even the use of wings, above the barnyard and its inhabitants. In a poem dominated by human allegory and metaphor, it is a slightly wistful reminder of an animal who escapes even the label of a particular species. At the same time, if the plumage of the rooster is a "vulgar beauty," that of the swallow is "gild[ed]": at least some of the difference is in the eye of the inevitably human observer. Amid the poem's more overt critiques of masculinity and fascism, Bishop also critiques these much smaller, subtler inconsistencies. And she does so throughout her work: this chapter has suggested that she returns again and again to the ways humans respond to and project onto animals of all kinds, both domesticated and wild. Her poems probe how we dote on and exploit them, train and blame them, pity and envy them, capture them in multiple senses—and how even the most careful human observers are susceptible to such impulses.

Notes

1. Auden, *Thank You, Fog*; from the personal library of Elizabeth Bishop, Houghton Library, Harvard University, call number *AC95.B5414.Zz974a.

2. An exception is in "North Haven," with its lists of flowers and reference to white-throated sparrows (*CP* 188). In her journal for those same years, the mid-1970s, Bishop writes of having acquired several guides to flora and fauna and writes that "I want now—now that it's too late—to learn the name of *everything*" (*North Haven Journal* 12).

3. The quoted words are from "The Prodigal" and "Florida," respectively.

4. See also "A Hen," one of the Clarice Lispector stories Bishop translated. It ends by making a point about human sentiment and selectiveness, which is likely one reason Bishop chose the story. The hen flees just before the routine Sunday slaughter; when she is finally recaptured, she suddenly lays an egg and begins to brood it, which causes the family to keep her as a pet rather than kill her. She lives on as a member of the household, "Until one day they killed her and ate her and the years went by" (*Prose* 386).

5. The word "blessed" also works as a kind of euphemistic oath, conveying the speaker's frustration and distress: she does not want to think about whaling.

6. Bishop recognized that many of us are fascinated by other species, although this fascination is not always in the other species' best interest. When she remembered visiting a circus with Marianne Moore at the start of their friendship, she saw even Moore as susceptible to a somewhat willful not-thinking-through of the exploitation: "I hate seeing animals in cages, especially small cages, and especially circus animals, but I think that Marianne, while probably feeling the same way, was so passionately interested in them, and knew so much about them, that she could put aside any pain or outrage for the time being" (*Prose* 120).

7. Bishop wrote delighted letters about Sammy: he "steals everything" and "eats six bananas a day. I must say they seem to go right through him & come out practically as good as new—meat, grapes—to see him swallowing grapes is rather like playing a pinball machine" (*One Art* 234). An earlier unfinished poem, drafted while he was alive, works as a list of his attributes: his beauty, combativeness, capacity for excitement, cleanliness, and so on (*EAPJB* 356–57).

8. For an overview of Descartes's belief that animals were essentially complicated automata, see Shugg.

9. The incongruity is helped by unexpected pentameter: "of the póor dúck the dóg decápitáted."

10. For real-life consequences of an aesthetic impulse, see "Crusoe in England," where Crusoe "dyed a baby goat bright red" "just to see / something a little different. / And then his mother wouldn't recognize him" (*CP* 165).

11. Bishop's keener perceptions tend to center on herself rather than on the racial others she sometimes includes in her texts; for a recent overview, see Parmar.

12. Carol Frost suggests the fish is an amberjack or a black drum (251–52); Ronald McFarland, that it is the red grouper (366). Bishop writes in her letters of catching a parrot fish (January 1939); in December 1948, she sends Lowell a postcard of a goliath grouper, with the words "These are the 'Fish'" (*WIA* 71).

13. Katie Van Wert notes the "deterioration" this description suggests (Hamil-

ton and Jones 33). In none of the species listed previously are such features of damage normal.

14. In the mid-1960s, when telling Anne Stevenson that Hemingway liked "The Fish," Bishop alluded to her changing views on blood sports: "[Hemingway] had the right idea about lots of things. (NOT about shooting animals. I used to like deep-sea fishing, too, and still go out once in a while, but without much pleasure, & in my younger tougher days I liked bull-fights, but I don't think I could sit through one now.)" (*Prose* 413).

15. "Roosters" is a good example of how Bishop's animals can invite such differing readings. For Hoorenman, Bishop "employs a complex syntax and diction that invoke a distinctly human realm, rather than a biologically more likely chicken-based perspective" (507); for Colm Tóibín, however, "'Roosters' is about roosters," specifically "roosters in Key West since roosters there dominate urban space as though they own it" (66).

16. This projection is similar to one in "Brazil, January 1, 1502," where male lizards watch "the smaller, female one, back-to, / her wicked tail straight up and over, / red as a red-hot wire" (*CP* 92). Who says the word "wicked"? It seems related both to the Spanish soldiers, in their moralism and their pursuit of indigenous women, and to the intensity of the actual lizards' attention (92).

17. MacRae notes that the birds are described with paint and other substances, suggesting "an attempt to make the rooster into a palpable artistic object; to pin down their otherness into a human frame of reference" (368).

Works Cited

Auden, W. H. *Thank You, Fog: Last Poems*. New York: Random House, 1974.

Bishop, Elizabeth. *The Complete Poems 1927–1979*. New York: Farrar, Straus and Giroux, 1983.

———. *Edgar Allan Poe and The Juke-Box*. Edited by Alice Quinn. New York: Farrar, Straus and Giroux, 2006.

———. *The North Haven Journal, 1974–1979*. Edited by Eleanor M. McPeck. North Haven, ME: North Haven Library, 2015.

———. *One Art: Letters*. New York: Farrar, Straus and Giroux, 1994.

———. *Prose*. Edited by Lloyd Schwartz. New York: Farrar, Straus and Giroux, 2011.

———. *Words in Air: The Complete Correspondence between Elizabeth Bishop and Robert Lowell*. Edited by Thomas Travisano, with Saskia Hamilton. New York: Farrar, Straus, and Giroux, 2008.

Cleghorn, Angus, ed. *Elizabeth Bishop and the Music of Literature*. Cham, Switzerland: Palgrave, 2019.

Cleghorn, Angus, and Jonathan Ellis, eds. *Elizabeth Bishop in Context*. Cambridge: Cambridge University Press, 2021.

Cook, Peter, Andrew Rouse, Margaret Wilson, and Colleen Reichmuth. "A California Sea Lion (*Zalophus californianus*) Can Keep the Beat: Motor Entrainment to Rhythmic Auditory Stimuli in a Non Vocal Mimic." *Journal of Comparative Psychology* 127, no. 4 (2013): 412–27.

Frost, Carol. "Elizabeth Bishop's Inner Eye." *New England Review* 25, no. 1/2 (2004): 250–57.

Furlani, Andre. "Elizabeth Bishop's Animal Manners." *Essays in Criticism* 70, no. 4 (October 2020): 428–46.

Hamilton, Geoff, and Brian Jones. *Encyclopedia of the Environment in American Literature*. Jefferson, NC: McFarland, 2013.

Hicok, Bethany. *Elizabeth Bishop's Brazil*. Charlottesville: University of Virginia Press, 2016.

Hoorenman, Johanna. "A *Lingua Unicornis*: Elizabeth Bishop and Anthropomorphism." *Contemporary Literature* 59, no. 4 (2018): 480–514.

Kalstone, David. *Becoming a Poet: Elizabeth Bishop, with Marianne Moore and Robert Lowell*. Ann Arbor: University of Michigan Press, 2001.

Komura, Toshiaki. "Poetics of Humility: Animal Ethics in Elizabeth Bishop and Robert Lowell." *Bishop–Lowell Studies* 2 (2022): 1–25.

MacRae, Marianne. "Animals." In *Elizabeth Bishop in Context*, edited by Angus Cleghorn and Jonathan Ellis, 359–70. Cambridge: Cambridge University Press, 2021.

Malay, Michael. *The Figure of the Animal in Modern and Contemporary Poetry*. London: Palgrave Macmillan, 2018.

McFarland, Ronald E. "Some Observations on Elizabeth Bishop's 'The Fish.'" *Arizona Quarterly* 38, no. 4 (1982): 365–76.

Oversight Hearing on Pfiesteria *and Its Impact on Our Fishery Resources*. October 9, 1997. Washington, DC: US Government Printing Office, 1998.

Parmar, Sandeep. "Race." In *Elizabeth Bishop in Context*, edited by Angus Cleghorn and Jonathan Ellis, 335–46. Cambridge: Cambridge University Press, 2021.

Paton, Priscilla. "'You Are Not Beppo': Elizabeth Bishop's Animals and Negotiation of Identity." *Mosaic: An Interdisciplinary Critical Journal* 39, no. 4 (2006): 197–213.

Pick, Anat. "Vulnerability." In *Critical Terms for Animal Studies*, edited by Lori Gruen, 410–23. Chicago: University of Chicago Press, 2018.

Schaefer, Dave. *Sailing to Hemingway's Cuba*. Dobbs Ferry, NY: Sheridan House, 2000.

Seiler, Claire. *Midcentury Suspension: Literature and Feeling in the Wake of World War II*. New York: Columbia University Press, 2020.

Shugg, Wallace. "The Cartesian Beast-Machine in English Literature (1663–1750)." *Journal of the History of Ideas* 29, no. 2 (1968): 279–92.

Spaide, Christopher. "Causes for Excess: Elizabeth Bishop's Eighty-Eight Exclamations." In *Elizabeth Bishop and the Music of Literature*, edited by Angus Cleghorn, 51–63. Cham, Switzerland: Palgrave, 2019.

Starbuck, George. "'The Work!': A Conversation with Elizabeth Bishop." *Ploughshares* 3, no. 3/4, issue 11 (Spring 1977): 11–29. https://www.pshares.org/issues/spring-1977/work-conversation-elizabeth-bishop.

Tóibín, Colm. *On Elizabeth Bishop*. Princeton, NJ: Princeton University Press, 2015.

Travisano, Thomas. "'A Very Important Violence of Tone': Bishop's 'Roosters' and Other Poems." In *Elizabeth Bishop and the Music of Literature*, edited by Angus Cleghorn, 65–78. Cham, Switzerland: Palgrave, 2019.

Turner, E. S. *All Heaven in a Rage*. New York: St. Martin's Press, 1965.

Walpole, S., and T. H. Huxley. "Disease among the Salmon of Many Rivers in England and Wales." *Bulletin of the United States Fish Commission*, vol. 1, for 1881. Washington, DC: Government Printing Office, 1882.

Witcher, Heather Bozant. "Archival Animals: Polyphonic Movement in Elizabeth Bishop's Drafts." In *Elizabeth Bishop and the Literary Archive*, edited by Bethany Hicok, 265–82. Ann Arbor: Lever Press, in partnership with Amherst College and Michigan Publishing, 2020. https://doi.org/10.3998/mpub.11649332.

CHAPTER 9

Learning from Animals in Yusef Komunyakaa's Poetry

DANIEL CROSS TURNER

Born James William Brown, the son of a carpenter, in the small mill town of Bogalusa, Louisiana, in the 1940s, Yusef Komunyakaa renamed himself to honor his grandfather, a stowaway on a ship to the United States from his native Trinidad. After graduating from high school, Komunyakaa joined the US Army and served a tour of duty as an information specialist during the American War in Vietnam from 1969 to 1970, where he saw active combat and was awarded a Bronze Star. In 1980, he earned an MFA in creative writing at the University of California at Irvine and afterward served on the faculty at the University of New Orleans, Indiana University, Princeton University, and New York University. To date, he has published more than fifteen books of poetry as well as two collections of essays and interviews. A remarkably wide-ranging writer, he has also adapted the ancient Sumerian epic *Gilgamesh* into a verse play, written the libretti of two operas, recorded three CDs of lyrical compilations with jazz musicians, and coedited two editions of a jazz poetry anthology. His status as a major contemporary poet has been recognized by literary critics and writers alike. To cite but a few of his numerous honors, Komunyakaa has been named a chancellor for the Academy of American Poets and elected a member of the Fellowship of Southern Writers, and his writing has garnered the Pulitzer Prize for Poetry, the Kingsley Tufts Poetry Award, the Hanes Poetry Prize, and the Ruth Lilly Poetry Prize. In 2005, *Callaloo*, a prominent journal of African American studies, dedicated a volume to Komunyakaa's life and work. Although the subjects he tackles in his verse are indeed extensive, he is most well-known for his poems exploring African American experience in the rural South and in the US War in Vietnam—two topics drawn from his personal history but that also connect to significant cultural moments.

Across Yusef Komunyakaa's prolific canon of poetry, animal imagery emerges as an insistent pattern that connects to substantial themes in his work. On one level, persistent figures of animals in his verse challenge ready divisions between human and nonhuman animals, questioning the basis for claiming clear human difference from, and therefore our dominance over, other creatures. Komunyakaa's animals undercut human-centered classifications that support the concept of anthropocentrism: that is, "the assumption of human superiority to other species and the denial of categories like language, morality, and thought to nonhuman animals" (Reynolds 157n1). His poetic creatures have a mind or consciousness and will of their own, often in opposition to, or in excess of, human designs and intentions. Deborah Clarke argues that thinking about animals in William Faulkner's fiction "allows us to recognize the limits of our ability to control and our ability to understand" (214). In a similar manner, animals in Komunyakaa's poems also disrupt "our assumptions of how things are ordered and structured" and therefore "allow us to think beyond the human" (214). In worlds created through Komunyakaa's verse, animals—human and nonhuman alike—undergo evolutionary development and flux, even crossing species lines. Sometimes humans transform into animals, or vice versa, yet often the transitions reflect more subtle evolutions evident over time. These latent and blatant human-animal metamorphoses tend to expose deeper similarities between human and nonhuman animals and trouble our assumptions of how things are structured, opening up fundamental questions about our traditional categories of meaning: What is human versus nonhuman? What is plant versus animal? What is natural versus unnatural?

Moreover, the world of animals, though frequently a place of wonder, is rarely placid. On the contrary, Komunyakaa's animals tend to be indifferent, recalcitrant, menacing, or violent; his beasts can be at once beautiful and brutal. Nature in Komunyakaa's poetry becomes "a source of curiosity, fear, and wonderment, as it simultaneously engulfs humans and is molded or consumed by us" (Salas 805). Human efforts to tame, suppress, or destroy animals usually backfire, often resulting in the humiliation of the human. Such moments force us to confront what Komunyakaa terms a "gut-level realism," a recognition of human fallibility against the "fearful certainty" of the nonhuman environs (Turner, "Remaking Myth" 343). When we deny our deep connection, even kinship, with animals in this manner, things end badly. If we damage animals and their surrounding environments for short-term human-centered gains, Komunyakaa warns that we may undermine our own ability to survive and thrive in the process, as we taint our own food sources and the natural ecologies that support human life as well.

On a further level, the recurrent motif of animals in Komunyakaa's poetry connects philosophical questions about the nature of animals to contemporary political contexts, revealing that modern civilization "has not eliminated violence, but casts it in new and less honest forms" (Kirsch 41). To cite lines from Komunyakaa's poem "The Four Evangelists," his verse creates a space where "we could / Almost talk with animals" (*Talking Dirty to the Gods* 51). We have much to learn from this poetic conversation, not only about the inner lives of animals and our connection with them but also about our current political engagements. Animal images in his work add to his poetry's criticism of damages inflicted by human misuse of the ecology, by military imperialism, and by racial inequity. In attending to these political themes, this chapter often looks to understand Komunyakaa's animals in specific locales, such as his hometown of Bogalusa, Louisiana, and Vietnam, where he was a soldier, in the process analyzing his poetic technique, particularly his expert attention to lineation.[1] To structure this chapter, I move more or less chronologically through Komunyakaa's life and oeuvre, shifting from his grappling with pollution's consequences in the rural South of his youth in "The Millpond" (1992); to his efforts to make sense of the beauty of the Vietnamese landscape amid the horror of its wartime destruction in *Dien Cai Dau* (1988); to his unexpected celebrations of the soundness and mathematic precision of maggots and slime molds in *Talking Dirty to the Gods* (2000); and, finally, to his embrace of hybridity while decrying human brutality in the art-poetry collaboration *Night Animals* (2019).

The Disturbing Lesson of "The Millpond"

Komunyakaa has long delved into matters of ecological justice in his verse using the lens of animality. Animal imagery is ubiquitous across his poetic accounts of the American South, especially as he exposes the effects of environmental damage. "The Millpond" is one of several poems set in and around Bogalusa that represent how humans can inflict damage on the animal world yet often wind up inflicting damage on the human realm in the process. The poem ultimately reveals how deeply interconnected various species really are, including how deeply connected the human food chain is to what we consider "lesser" creatures. The poem also captures the feeling of "jagged symmetry" at the heart of the natural world that Komunyakaa described in an interview with the author: "I liked going out into the woods at five or six years old, and everything seemed so immense and so mysterious to me. It made my senses come to life. I felt that there was a nervous edge to the world, to human existence. That it wasn't neatly tied up in a gift box. There was a kind of jagged symmetry to everything" (Turner, "Remaking Myth" 338). "The Millpond" reflects the sense Komunyakaa felt as a young boy in Bogalusa that the natural world held a deep attraction for him but was also a source of uneasiness, even threat. The poem begins, innocently enough, with a glimpse of an inspiring pastoral scene—evoking a literary genre in which rural life and settings are idealized—as swamp orchids are seen to shimmer on the pond's surface: "They looked like wood ibis / From a distance" (*Pleasure Dome* 270). From the start, however, there are hints of "jagged" edges among the ostensible "symmetry" of the opening vista; something about this place feels off-center, knotty, fragile.

As the speaker draws closer, he sees that the orchids "became knots left for gods / To undo, like bows tied / At the center of weakness" (270). The teeming, murky millpond promises things that appear to the speaker's young eyes at once divine and devious, attractive and frightening: "Gods lived under that mud / When I was young & sublimely / Blind" (271), for "Each bloom" offered "a shudder of uneasiness" (271). As he grows older and less sublimely blind, he begins to see that there is something unnatural about the place. The millpond is filled with slag water, toxic runoff from the local paper mill in Bogalusa. The interdependence of human and animal is suggested with the pointed image of "slate-blue catfish / Headed for some boy's hook / On the other side" (270). An instance of boyhood innocence—catching some catfish at the local pond—transforms to mature revulsion when we realize the catastrophic effects of the mill's poison runoff on the surrounding ecology: "five-eyed / Fish with milky bones / Flip-flopped in oily grass" (271). The fish have mutated through the "chemical water" (271), the continual toxic slurry from the mill. The grotesque image of deformed fish out of water is followed by other creatures inhabiting the millpond ecosystem that also face a powerful ecological threat, symbolized by the "reflection of a smokestack" that "Cut[s] the black water in half" (271). Snapping turtles, mosquitoes, and bullfrogs "Singing a cruel happiness" (271) are reduced to "decoys / For some greater bounty" (271). The potentially fatal interdependence of humans and animals is underscored by the poem's closing image of the young speaker capturing crawfish from the tainted pond with salt meat tied to nylon string:

They clung
To desire, like the times
I clutched something dangerous
& couldn't let go. (272)

Like the crawfish, we humans, in the context of long-term ecological sustainability, also seem trapped by our own limited, immediate desire, in spite of the danger to our own existence. "The Millpond" suggests this imbalanced condition ultimately is not viable. We must learn to let go of our limiting anthropocentric desire to overpower the ecology for short-term profits, for in tainting the groundwater, we taint intertwining ecosystems, including our own food sources, and therefore we taint ourselves. A poem that begins with a brief hint of the pastoral ends with a strong antipastoral critique of how waste from the paper mill—Bogalusa's economic lifeblood—is poisoning the community's ecological bloodlines: plant, animal, and human.

Beauty and Horror in the Vietnam Poems

Donovan McAbee links two major concerns of Komunyakaa's work—ecology and race—in contending that the stark indifference or recalcitrance of nature in Komunyakaa's vision inspires resistance to racist constructions of identity.[2] These themes are evident as animals appear throughout Komunyakaa's poems about his time as an African American army officer in the US War in Vietnam during the late 1960s. His best-known poetic accounts of the Vietnam War are concentrated in the volume *Dien Cai Dau*. The title comes from the Vietnamese term meaning "crazy in the head," which was a phrase often applied to American soldiers by the native people. Komunyakaa's Vietnam poetry, according to Herman Beavers, moves toward "complicating our understanding of the war" by insisting that "there is something beyond destruction and tragedy, in a manner that eradicates the hegemony of scale" (357). Eradicating the hegemony of scale in the context of Komunyakaa's animal imagery means humans no longer hold pride of place atop a great chain of being; instead, we must come to terms with our physical limits and the mortality we share with other animals. In so doing, we further need to acknowledge the severe damage our conflicts inflict on nonhuman creatures and the ecologies we cohabit through powerfully destructive weaponry. The traumatic setting of the Vietnam War foregrounds this awareness of human frailty and reveals our material closeness to animals. In an interview with Muna Asali, Komunyakaa interestingly claimed that while on combat tour in Vietnam he "did not fear the land" and even "realized a kind of beauty in the overall landscape," though this beauty was interwoven with violence (78). In his Vietnam-based poems, the American military struggles to adapt to the local ecology, especially its wildlife, which pushes back against the presence of the occupying troops (fig. 9.1). The Vietnamese terrain, with its surfeit of wild animals, can be perilous, treacherous, yet at the same time it inspires astounding beauty in Komunyakaa's verse.

As the title of a representative poem suggests, the jungle fosters "A Greenness Taller Than Gods." The sheer force of the nonhuman environment, embodied in its sublime "greenness," channels an overwhelming power that seems to exceed the divine ("taller than gods"). The poem's opening lines are freighted with animality. In contrast to the stasis of the US soldiers ("When we stop"), the jungle teems with animal motion and sound:

> *When we stop,*
> *a green snake starts again*
> *through deep branches.*
> *Spiders mend webs we marched into.*
> *Monkeys jabber in flame trees,*
> *dancing on the limbs to make*
> *fire-colored petals fall. Torch birds*
> *burn through the dark-green day. (Pleasure Dome 196)*

The animals are simultaneously threatened by and threatening to the invading soldiers. Figures of height and depth suggest the dense power of the surrounding ecology. Bright and layered color imagery attached to the animals ("green," "flame," "fire-colored," "dark-green") strengthens their allure, even as they radiate warning, or menace. The green snake begins the natural world's movements, which gather momentum through intricate designs of spiders mending webs and monkeys jabbering and dancing in flame trees while making "fire-colored petals fall," and finally resolve in a dynamic visual flash: "Torch birds / burn through the dark-green day." Each animal exudes beauty in itself yet also carries underlying peril for the invasive human presence in its midst. The green snake evokes the unruly serpent in Eden and its animus toward the human, while the spider webs momentarily disrupt the soldiers' march. The fire-colored petals released by the monkeys' heightened activity in the trees puts us in mind of searing bomb strikes descending from on high, while torch birds become the ecology's figurative equivalent of napalm's unending burn. All told, this amalgamation of animals creates for the American soldiers a hostile, if sublime, environment.

The excessive materiality of the Vietnamese forest and its animal life is contrasted with the abstraction of military topography, as a "lieutenant puts on sunglasses / & points to an *X* circled / on his map" (196–97). The lieutenant's sunglasses reflect his skewed, protected vision of the surrounding ecology, when he enters "a clearing that blinds" (197). The soldiers are dangerously out of sync with nature. To better survive in such a dense, unfamiliar, and fraught landscape, they need to "learn / to move like trees move"

Figure 9.1. Two Tonkin snub-nosed monkeys (*Rhinopithecus avunculus*) in northeastern Vietnam. Photo by Quyet Le. Wikimedia Commons (CC BY-SA 2.0).

(197). US troops narrowly avoid a surprise encounter with a Viet Cong outfit, signified by their lingering trace in "branches left quivering" (197). Set against the immense Vietnamese ecology, the unnerving depth and height of its greenness and animating life force, the Americans seem in danger of becoming a mere shadowy presence:

> *We move like a platoon of silhouettes*
> *balancing sledge hammers on our heads,*
> *unaware our shadows have untied*
> *from us, wandered off*
> *& gotten lost. (197)*

Confronted with the physical presence of the wild, vibrant animals surrounding them, the US soldiers wither to ghosts.

In other places, we witness lasting damages US military operations have imposed on the Vietnamese terrain. In "The Edge," the native Vietnamese women's trauma mirrors that of the landscape, as they endure "ragged dreams / fire has singed the edges of" like the "slow dying the fields have come to terms with" after years of warfare (*Pleasure Dome* 212). "Toys in a Field" is loaded with animalistic imagery, through which Vietnamese children take the place of native animals in a posttraumatic ecology. The children create a makeshift playground from the destroyed "toys" left behind by American troops. "Using the gun mounts / for monkey bars," the children play "skin the cat" on the ruins "of abandoned helicopters / in graveyards" (133). In a further layering of animalistic imagery, the children figuratively manifest as birds: "With arms / spread-eagled they imitate / vultures landing in fields" (133). The promise of liberation by US forces, suggested in the eagle image, has yielded to the grim postwar reality of feeding on dead remains, as vultures. The children's play is eerily silent "as distant rain / the volume turned down / on the 6 o'clock news" (133), suggesting how the Vietnamese find themselves abandoned—a wounded ecology reflecting a wounded nation—now that the volume of attention to the war on American televisions has diminished. At the poem's close, the silence is broken by an animalistic sound that marks another postwar inheritance from the US military presence in Vietnam: the haunting figure of a "boy / with American eyes" who "keeps singing / rat-a-tat-tat, hugging / a broken machine gun" (133). The Amerasian boy, presumably left behind by his American soldier-father, reflects another broken thing abandoned among the scarred landscape, and his uncanny song and derelict weapon signal that the cycle of violence will likely continue in the wake of the US military's exit from Vietnam. The motif of animals coursing through Komunyakaa's Vietnam poetry points to US soldiers as an unnatural presence in the native terrain and suggests that American military imperialism is likewise perilously out of place in Southeast Asia, leaving in its wake destruction and trauma.

Celebrating Overlooked Creatures in *Talking Dirty to the Gods*

Animal imagery also forms a central motif in *Talking Dirty to the Gods*, Komunyakaa's most formally structured verse collection. The volume contains "132 poems of four unrhymed quatrains each," with most lines in each quatrain containing "four syncopated beats," producing "the tensions of a four-by-four-by-four framework" (Salas 799). These four-by-four-by-four poems are "sonnetlike in their contained, patterned arguments and in their collective narrative impact" (36), creating a "tight form and epigrammatic style" (Bernard 36). Like the poetic equivalent of an infinity mirror, *Talking Dirty to the Gods* presents a vast catalog of figures and allusions, as Komunyakaa "has populated his poems with sometimes frenetic allusions to Greek, Roman, Christian, Japanese, and every other mythology going. At times, the book is reminiscent of a Walt Disney cartoon in its cavalier, knowing collision of cultural artifacts: here the seven deadly sins, Bellerophon, Romeo and Juliet, Shiva, an incubus, Chet Baker, and Sleeping Beauty all talk to us or to one another, sometimes in the same poem" (Bernard 36). Komunyakaa's creative "associational play" from diverse image to image produces "the oddball but exact quality of surrealism at its best" (Wojahn 169), while the volume's short lines and tight rhythms lend "a montage-like pacing to these effects" (169).

Most notable among the myriad allusions and images streaming through *Talking Dirty to the Gods* are figures of animals, which often eat away at anthropocentric divisions between human and nonhuman creatures. "Ode to the Maggot" offers a tour de force sampling of Komunyakaa's intense form (fig. 9.2). The poem's dynamic compression echoes the minuscule yet fearsome power of its title creature, as Komunyakaa celebrates the irrepressible force of this unseemly insect to "Go to the root of all things" (*Talking Dirty to the Gods* 10). In breaking down organic forms to their essential elements, the maggot reveals the basic, inexorable process underlying nature to be "sound & mathematical" (10): the maggot builds off destruction, thriving by way of putrefaction. This tiny "Brother of the blowfly" is "merciless / With the truth" (10)—namely, the blunt reality of the natural process of death, which in turn gives life to the maggot. Decay's totemic animal is honored as the "Little / Master of earth" that can "take every living thing apart" (10), utterly indifferent to divisions between human and nonhuman animals, to artificial hierarchies of social status, for the maggot can equally "cast spells on beggars & kings" (10). The poem's taut arrangement, its four-by-four-by-four structure, dovetails with its subject and theme: this minute, typically overlooked animal in fact represents a microcosm of the organic process of death and life.

"Slime Molds" follows "Ode to the Maggot" as a companionate piece, expanding Komunyakaa's consideration of the nature of animals by addressing taxonomic classifications that categorize organisms into various kingdoms as animals, plants, or fungi. As he does with the maggot, Komunyakaa singles out another curious yet deceptively intricate and powerful creature, the slime mold, for poetic exploration and wry celebration (fig. 9.3). Like the maggot, slime molds represent a somewhat distressing presence from a human viewpoint, since they emerge out of nowhere and are considered eyesores; slime molds, for instance, are commonly said to resemble "dog vomit." Yet slime molds' actions are remarkably, to borrow a phrase, "sound & mathematical." We don't know quite what to make of slime molds or quite what they're after. From the perspective of modern biology, slime molds are naturally occurring hybrids, since they exhibit taxonomic traits of at least three commonly recognized kingdoms: fungi (Fungi), plants (Plantae), and animals (Animalia).[3] Yet slime molds move forward with an inexorable, calculated will. Slime molds cause confusion among us, while they themselves seem to suffer none.

Reflecting the trademark compression of *Talking Dirty to the Gods*, "Slime Molds" opens with a terse, two-word sentence that succinctly introduces the uncanny presence of this indefinable creature: "They're here" (11). This is the stuff of horror movies. The opening line seems to warn of alien invasion. Indeed, the *New York Times* published a couple of articles in May 1973 about a Dallas, Texas, housewife who, concerned she might be witnessing the presence of alien life-forms or "blobs," reported yellow slime molds slinking across her backyard ("Blobs"; "Texas Scientists"). The poem's

Figure 9.2. Decomposing Virginia opossum (*Didelphis virginiana*), showing maggots. Photo by Tim Vickers. Wikimedia Commons.

opening salvo forces us to confront the uncomfortable unavoidability of these odd things, the reality of these seemingly unreal creatures. We must face not merely their repulsiveness but also their resistance to our systems of classification, to our powers of scientific understanding. We see slime molds emerge "Among blades / Of grass" (11), an allusion to Walt Whitman's bullish heralding of American democracy in *Leaves of Grass* (1855). For Whitman, the individual blade seamlessly integrates into the healthy, common field of grass, as the particular US citizen contributes to a thriving democratic culture. On a political level, Komunyakaa's poem undercuts Whitman's optimism somewhat, as "spores / Glom together" into a single organism. If, like Whitman, Komunyakaa figures the American democratic mass as an organic unity, the slime mold metaphor deromanticizes Whitman's vision of democracy as a productive natural whole. If the slime mold unifies parts into a whole to work together, it does so toward what purpose? It is "Good / For nothing" (11). Its status as organic whole is undermined as the slime mold is described in terms of a human construction: "Pieces of a puzzle" (11). Even unified, the slime mold remains "scrambled," a bit revolting, still "like divided cells" (11). In the slime mold, we see the messiness of American democracy, its intransigence and penchant for division, thus questioning Whitman's exuberant celebration of the US democratic project.

And then "Slime Molds" makes an abrupt turn, one of Komunyakaa's unexpected associative shifts, as the poem jump-cuts from the political to the philosophical. Slime molds are said to be "Something / Left over from a world before— / Beyond modern reason" (11). This reference to "a world before" suggests that ecological deep time is still with us, not locked away in a primordial past. If, in contrast to more "noble" totemic animals (the bear, the wolf, the lion, for instance), the slime mold appears uninspiring, this hybrid organism speaks directly to primeval conditions that remain with us and underpin many of the abstract structures of modernity. Slime molds are "Beyond modern reason," evading our typical categories of classification, lingering, in sheer ambiguity, somewhere between science and science fiction. Further, as it aggregates into a single working organism, the slime mold consolidates into "Primeval / Fingers reduced & multiplied / A hundredfold" (11); it is one and yet many, reduced and multiplied simultaneously. The component organisms have surrendered to the whole because "the most basic / Love & need shaped them into a belief / System" (11). The individual organisms join the larger framework to seek food and light and growth. However, beyond biological need, Komunyakaa adds the notion of love, of an emotive connection with others, an instinctual attraction between bodies that causes them to work together in a collaborative belief system as powerfully motivating as the needs of raw survival. Slime molds are social animals/plants. They organize intelligently for common action. If democracy on the large scale is messy and fractious, then kinship on the smaller scale—what the poem calls "love"—might be an essential part of animal life. And so the slime mold becomes "Good for something we never thought / About" (11). Yes, slime molds are still truculent things, still uncanny, alienating presences on earth: "these pets of aliens crawl up / The Judas trees in bloom" (11). However, in a deeper sense, slime molds offer us an inspiring model for collective action and also are good for troubling human-centered hierarchies for the natural world, teaching us that the things of nature are themselves unstable, in flux, including humans. "Slime Molds"—and *Talking Dirty to the Gods* generally—thus embraces an ethos of evolutionary transformation, of biological metamorphosis continuously in the making.

Figure 9.3. *Slime mould*. Photo by Bernard Spragg. Wikimedia Commons.

Embracing Hybridity and Decrying Human Brutality in *Night Animals*

Komunyakaa's most recent foray into animality takes this process even further. Arguably his most thorough examination of the inner lives of animals, *Night Animals* is a brilliant collaboration with visual artist Rachel Bliss, whose stunning images accompany Komunyakaa's poems. Bliss's paintings mirror Komunyakaa's surrealistic technique, interlacing animal with human forms to create unsettling human-animal hybrids that blur traditional boundary lines between humans and other creatures, raising the question of where humanity ends and animality begins. Bliss imbues animal-like bodies with humanoid features—particularly eyes—and then reverses the procedure: her humanlike figures are marked with animalistic traits, like whiskers, fur, fangs, eye masks, and eyespots (fig. 9.4). Although certain identifiable creatures are suggested by Bliss's images (e.g., the mockingbird, cat, flying fox, platypus, armadillo, Tasmanian devil), which are portrayed in particular poems by Komunyakaa, the overall affect is of a dreamlike, at times nightmarish, suggestiveness, but one grounded in pinpoint perception of reality. If chimerical in their syncretic forms, these hybrid creatures announce themselves as very real presences; they are deeply expressive—each reflects a distinct personality—and demand our attention.

Bliss's images combined with Komunyakaa's poetic interpretations point to permeable boundaries between human and nonhuman animals, underscored by a motif of liminality through the volume, of things on the verge. The creatures Bliss and Komunyakaa depict form odd composites, studies in ongoing species adaptation and evolutionary biology. Through Bliss's arresting images and Komunyakaa's piquant, probing verse explorations, we encounter an array of strange creatures that stretch our typical sense of what's natural. We meet mockingbirds who not only mimic other birds' songs but mix in "a few human words" (*Night Animals* 1); airborne foxes with wings and "bellies / pale & tight as bloated drums" (7) who are "masters of drunk acrobatics" (7); an armadillo who seems a natural-artificial amalgam, a "little suitcase of guts & nails" with "broken hinges" (23); and a venomous platypus who reflects an uncanny metamorphosis-in-process as part duck, part beaver, part otter, a vestige from "an old world, / a prototype, the first chimera / pieced together by a prankish god" (9).

Across such combining images and words, *Night Animals* reveals how the world of animals eerily surpasses "nature" in the traditional anthropomorphic sense of *natura naturata*, the nonhuman environment as "passive matter organized into an eternal order of Creation" (Bennett 117). According to *natura naturata*, only humans possess the capacity for reason and speech, and therefore we humans should exert supreme power over the natural world and use its creatures entirely to our benefit. In this view, animals are essentially machines that possess no consciousness or inner life and are therefore meant to be under human control. Komunyakaa's creature-filled settings, coupled with the hybrid figures of Bliss's art, instead incarnate *natura natarans*: "the uncaused causality that ceaselessly generates new forms" (Bennett 117), a "process of collaboration and contestation between bodies" that is "not random or unstructured, but conforms to the strange logic of vortices, spirals, and eddies" (118). The world of *natura natarans* is not made up of static materials and inert beings but is full of kinetic matter and vibrant creatures with the capacity of thought and language, as well as ethical choice and action. In other words, according to Bennett, *natura natarans* represents a profoundly non-anthropocentric realm, where humans are no longer viewed as apart and above the rest of nature.[4]

Bliss and Komunyakaa's artistic-poetic collection portrays animals as integral elements in the complex, ever-developing systems of *natura natarans*. This idea merges in *Night Animals* with the notion that we humans can no longer assume our central place over nature and our unquestioned command over animals. Humans are not the only animals who create cultures, and we can learn organization and other valuable social behaviors and emotions from animals. In condescending anthropomorphic fashion, we often assign negative emotions to animals, blaming them for our base instincts, including a penchant for predatory violence and sexuality. However, humans more often do violence to animals, rather than the other way around; we trap, domesticate, or kill and eat animals, and we also destroy them by proxy in our wars.[5]

In "The Leopard" from *Night Animals*, Komunyakaa reveals nature's brute force, while reveling in its sublime beauty through the ultra-sleek figure of the leopardess, who takes the form of "a goddess in a world mastered / by repetition & unearthly cadence" (10). Out tracking prey in the night, the leopardess, in pure certainty, is "pacing off light hidden in darkness" (10), and her instinctual movements are like "coming to an answer / of the oldest unspoken question" (10). In this game of predator and prey, of kill or be killed, there is no law, except that of the jungle. The sheer power of the leopard's singular graceful violence is beyond human hierarchies of good and evil: "The mitigated laws of kingdom, / district, & tribe do not matter here" (10). Like the impressive big cat featured in William Blake's "The Tyger" (1794), Komunyakaa's leopardess also burns bright in the forests of the night with a shimmering capacity for quick and fatal action, and we likewise wonder what, if any, immortal presence could dare frame such fearful symmetry (fig. 9.5). Her dauntless, unfaltering night stalking embodies "a wild, simple knowledge" (10), rooted purely in "Sinew, muscle, gratitude" (10), as the leopardess prepares to "ride another animal / down to growl, tussle, gristle, / & blood-lit veins" (10). So it goes, the cycle of brute nature.

The takeaway from "The Leopard," one might suppose, is that a primal instinct for dominance is a basic underpinning of all life. The closing image, however, resists a facile conclusion. Following the bloody scene of predation, we are left hanging with the ambiguous figure of "leaves left / quivering in the passing night" (10). The leaves are still "quivering" from the intense violence, signaling the resistance of the surrounding environment, and the night, the time when the leopard hunts, is "passing." A companion

Figure 9.4. *Tamara* (2012), by Rachel Bliss. *Night Animals* (Louisville, KY: 2020).

poem, "Snow Tiger," follows "The Leopard" in *Night Animals* and invokes the figure of another female big cat, but this time transfiguring violence into empathy, since "There's always a mother / of some other creature / born to fight for her young" (11). The snow tiger's penchant for protecting and nurturing her young makes us question "if hunger is the only passion" (11). The dynamic of predator versus prey is a crucial facet of the animal world, yet it is not the only important element. Other passions beyond hunger motivate animals, including the structuring power of caring for young and building complex family and social arrangements to protect and nurture offspring. Nature, in Komunyakaa's telling, is a process of collaboration as well as contestation with its own logic and agency.

"Nightriders" illustrates the collision between animal and human cultures in *Night Animals*, as the nonhuman ecology of the American South becomes an unwilling repository for human lives cut short through the terrors of racialized lynching. The environment itself recoils against the dark human brutality buried within it, as an otherwise stable ecological balance reacts to the violent intrusion of lynch law, for "The fleshy scent of magnolia rekindles the years, / & a single night cannibalizes a century of blooms" (21). The nonhuman surroundings and their animal inhabitants disavow this human savagery, symbolized by a hoot owl that recalls names of African American victims of white supremacist hatred: "In a deep vista of scrub oak a hoot owl names / the lost ones, growing into the never-heard-of" (21). The poem's concluding image makes it clear that lynching's time was always out of date, an ugly medieval practice based on false codes of modern "chivalry" as cruel as they are self-defeating: "dogwood / branches sag, & ugliest of the four horsemen / languishes in medieval garb, feet in a noose" (21). The sagging dogwood branches show how the natural world pushes back against racist violence committed in its midst. The "ugliest of the four horsemen" in his "medieval garb" recalls the "knights" of the Ku Klux Klan in their similarly archaic robes and hoods as they terrorized African American victims on apocalyptic night rides. "Nightriders" suggests that the cycle of racist violence was self-defeating for the white nightriders; as they inflicted brutality on African Americans, white participants became entangled in self-destructive hatred, symbolized by the horseman's own "feet in a noose."

Like "Nightriders," "Another Kind of Night" draws out intersections between nonhuman and human cultures, unveiling a shared history of violence but also, perhaps surprisingly, of mercy. Where "Nightriders" uses the nonhuman world to critique white southern lynch law, "Another Kind of Night" invokes animal parallels to re-create the horrors of the Middle Passage, where millions of Africans perished in subhuman conditions while being transported across the Atlantic on European slave ships. In the bleak holds of these ships, enslaved Africans are stripped of individual identities to such an extent that they metamorphose grimly into an indistinguishable animalistic mass: "down there where everything is one // godforsaken animal wounded in the nameless dark" (19). Thus deindividualized, the Africans are viewed by their European captors as devoid of human language, as a fearsome inarticulate beast to be controlled by force, for the captives are "Moans in the belly of a writhing leviathan" (19). In this unnatural kind of "night," the enslaved are marked as living dead, as a troubling undead presence that disrupts the false ideals of a European colonial civilization based on chattel slave labor: "All the faces are one constellation where the dead / interrogate the living" (19). Under such conditions, where one set of humans treats another set as "nothing / but cargo" (19), the poem turns attention to the animal world to seek a better way. Komunyakaa intersperses "Another Kind of Night" with a litany of images of African flora and fauna, all with positive implications of stability, beauty, and peace, such as "remembered trees, a nine-hundred-year-old sky, / the call of birds, a mamba snake snoozing beside a stone" (19). In addition, we see "jackals holding a ceremony at the edge of a lake" (19), revealing the social organization of animals and showing that ritual is not uniquely human. Unfortunately, the enslaved Africans' memory of these animals "all fades into rancor" (19) against the pressing inhumane conditions they endure inside the slave vessels. Against the nightmare of this ironic "new world" (19), the abuses wrought by modern human "civilization," we are asked to take a lesson from, of all creatures, the hyena, an animal we typically single out for contempt: "Hyenas on a hilltop, / are you still trying to tell me something about mercy"? (19). Hyenas are not known for showing mercy. Compared to the coordinated, intentional acts of savagery committed by the slavers, in a biting reversal, perhaps we could learn something about mercy even from hyenas, after all.

In closing, animals make up a vital, insistent motif throughout Yusef Komunyakaa's poetry. Images of animals often serve to critique the excesses of human behavior, especially our actions that inflict violence on other humans and nonhuman creatures and damage the ecology. Furthermore, Komunyakaa reveals that animals are not just hollow machines but are capable of thought, intention, will, and resilience. Animals can even offer us at times a potential model for positive ethics through collective action and a sense of nurture and empathy, perhaps even mercy. In the end, Komunyakaa's verse illustrates some of the many ways we can, and should, learn from other animals.

Notes

1. Several critics have analyzed Komunyakaa's poetics in relation to jazz, focusing on a shared emphasis on improvisation between jazz and Komunyakaa's metrical variations. Keith Leonard argues that Komunyakaa's jazz-influenced improvisations critique

Figure 9.5. Female leopard (*Panthera pardus*) descending from its favorite tree, where it spent the warmest hours of the day. Londolozi, Sabi Sand, South Africa. Arturo de Frias Marques. Wikimedia Commons (CC BY-SA 4.0).

racist definitions of identity, representing "a postmodern introspective practice that rewrites the social discourses that create and justify exclusion, including but not limited to racism, making it the defining activity of the mind" (826). Meta DuEwa Jones offers an intriguing model for contemporary African American jazz poetry that accounts for "its intricately interwoven visual and aural contours" (87), while Keith Cartwright merges Komunyakaa's jazz poetics with the Haitian Creole notion of *balanse*: the Vodou ritual of enlivening by rhythmically evoking conflict and contradiction, creating a "mojo-blues ethos" (863). In *Southern Crossings: Poetry, Memory, and the Transcultural South*, I interpret Komunyakaa's improvisational technique in light of a poetics of trauma, grounded in the American South and Vietnam.

2. In "Modern Metamorphoses and the Primal Sublime," an article on poems by Komunyakaa and by Derek Walcott set in the American South or the Caribbean, I argue the motif of human-to-animal metamorphosis in these poets' work enhances "our understanding of human injustices and of the value of nonhuman nature" (53). Such metamorphoses not only reveal the intimate kinship between humans and animals but

also challenge "socio-cultural forms of mastery that inform the South's and the Caribbean's shared histories of colonization, enslavement, and apartheid" (53).

3. Once grouped under kingdom Fungi because they produce abundant spores, slime molds' taxonomic classification grew uncertain under closer scrutiny. Like plants, slime molds contain cellulose within their cell walls, yet like animals, they are capable of coordinated movement, using chemical signals from food sources to detect the easiest direction to reach nutrients. A particular slime mold, *Dictyosteliida*, is a large, undivided cell, but when it secretes a specific chemical, individual amoebae unite and work together as a single organism. Their process of cellular communication and aggregation has been mapped with mathematical equations so precise that these have been used as the basis for computer programs and video games. Because slime molds have proven so resistant to taxonomic categories, they are now classified under a separate "catch-all" kingdom, Protista, which contains all eukaryotes (i.e., organisms made up of cells containing a membrane-bound nucleus) that are not classified as fungi, plants, or animals.

4. Bennett is one of several recent theorists, such as Claire Colebrook, Bruno Latour, and Timothy Morton, who espouse a similar notion of the nonhuman ecology and are associated with object-oriented ontology or the new materialisms. As Anthony Reynolds notes, however, this line of thought has deep roots. In his analysis of Jack London's writing, Reynolds unpacks nineteenth-century US philosopher Charles Sanders Peirce's view that thought was "an emergent property of matter itself and thus distributed throughout the natural world" (139), noting that this "attribution of the quality of mind to matter, known as panpsychism, has a long history dating back to the pre-Socratics and has become prevalent again more recently in the various theoretical movements devoted to material agency that have come to be known collectively as the new materialisms" (139).

5. Komunyakaa gives the link between human and nonhuman animals a further turn by exposing how modernity often sugarcoats a dark will to power and desire to dominate, as in the figure of the corporate tycoon as shaved monkey in "Hearsay" from *Talking Dirty to the Gods*:

> *Yes, they say if you shave a monkey*
> *You'll find a pragmatist, the president*
> *Of a munitions plant, a tobacco tycoon,*
> *Or a manufacturer of silicon breasts (3)*

The figure of the Darwinian corporate executive as a trumped-up primate willing to sell anything to anyone, no matter how harmful, when the price is right, is made especially grim with the mention of a munitions plant—an image that connects to Komunyakaa's rich corpus of war poetry.

Works Cited

Asali, Muna. "An Interview with Yusef Komunyakaa." In *Blue Notes: Essays, Interviews, and Commentaries*, edited by Radiclani Clytus, 76–84. Ann Arbor: University of Michigan Press, 2000.

Beavers, Herman. "Till the Hurt Becomes Music: Gnosticism and Improvisation in the Poetry of Yusef Komunyakaa." In *A History of the Literature of the U.S. South*, edited by Harilaos Stecopoulous, 342–61. New York: Cambridge University Press, 2021.

Bennett, Jane. *Vibrant Matter: A Political Ecology of Things*. Durham, NC: Duke University Press, 2010.

Bernard, April. "I Sing of Slime Mold." Review of *Talking Dirty to the Gods*. *New York Times Book Review*, December 2000, 36.

"'Blob' Grows in Dallas; Housewife Is Puzzled." *New York Times*, May 30, 1973, 14. https://www.nytimes.com/1973/05/30/archives/blob-grows-in-dallas-housewife-is-puzzled.html.

Cartwright, Keith. "Weave a Circle round Him Thrice: Komunyakaa's Hoodoo Balancing Act." *Callaloo* 28, no. 3 (Summer 2005): 851–63.

Clarke, Deborah. "Faulkner's Animals: Testing the Limits of the Human." In *Animals in the American Classics: How Natural History Inspired Great Fiction*, edited by John Cullen Gruesser, 199–215. College Station: Texas A&M University Press, 2022.

Colebrook, Claire. "Queer Vitalism." *New Formations* 68 (2010): 77–92.

Jones, Meta DuEwa. "Jazz Prosodies: Orality and Textuality." *Callaloo* 25, no. 1 (2002): 66–91.

Kirsch, Adam. "Verse Averse." Review of *Talking Dirty to the Gods*. *New Republic*, February 26, 2001, 38–41.

Komunyakaa, Yusef. *Pleasure Dome: New and Collected Poems*. Middletown, CT: Wesleyan University Press, 2001.

———. *Talking Dirty to the Gods*. New York: Farrar, Straus and Giroux, 2001.

Komunyakaa, Yusef, and Rachel Bliss. *Night Animals*. Louisville, KY: Sarabande, 2019.

Latour, Bruno. *We Have Never Been Modern*. Translated by Catherine Porter. Cambridge, MA: Harvard University Press, 1993.

Leonard, Keith. "Yusef Komunyakaa's Blues: The Postmodern Music of *Neon Vernacular*." *Callaloo* 28, no. 3 (2005): 825–49.

McAbee, Donovan. "'Song from the Soil': The Role of Nature in Yusef Komunyakaa's Poetry." *English Studies: A Journal of English Language and Literature* 100, no. 4 (2019): 447–60.

Morton, Timothy. *Hyperobjects: Philosophy and Ecology after the End of the World*. Minneapolis: University of Minnesota Press, 2013.

Reynolds, Anthony. "Learning to Think like an Animal: Pragmatism in Jack London's *The Call of the Wild*." In *Animals in the American Classics: How Natural History Inspired Great Fiction*, edited by John Cullen Gruesser, 133–60. College Station: Texas A&M University Press, 2022.

Salas, Angela M. "*Talking Dirty to the Gods* and the Infinitude of Language: Or Mr. Komunyakaa's Cabinet of Wonder." *Callaloo* 28, no. 3 (Summer 2005): 798–811.

"Texas Scientists Think Backyard Blob Is Dead." *New York Times*, May 31, 1973, 82. https://www.nytimes.com/1973/05/31/archives/texas-scientists-think-backyard-blob-is-dead.html.

Turner, Daniel Cross. "Modern Metamorphoses and the Primal Sublime: The Southern/Caribbean Poetry of Yusef Komunyakaa and Derek Walcott." In "The South and The Sublime," special issue, *Southern Quarterly* 48, no. 3 (2011): 52–69.

———. "Remaking Myth in Yusef Komunyakaa's *Talking Dirty to the Gods, Taboo*, and *Gilgamesh*: An Interview." *Mississippi Quarterly: The Journal of Southern Cultures* 62, no. 2 (2009): 335–50.

———. *Southern Crossings: Poetry, Memory, and the Transcultural South*. Knoxville: University of Tennessee Press, 2012.

Wojahn, David. Review of *Pleasure Dome: New and Collected Poems / Talking Dirty to the Gods*. *Poetry* 179, no. 3 (December 2001): 168–72.

CHAPTER 10

Cawing in the Dark
Avian Alterity in the Poetry of Joy Harjo

THOMAS C. GANNON

On June 19, 2019, Mvskoke poet Joy Harjo was appointed Poet Laureate of the United States. In a crucial sense, Harjo is the *perfect* choice as the first Native US Laureate because her work skillfully balances two tonalities in contemporary Native literature: on the one hand, edgy, even scathing, in her presentations of Native life in a still racist contemporary America yet, on the other hand, positively visionary, with an ultimate message of love and hope and transformation—for herself, her people, and her planet.

In terms of the planet, Harjo's laureateship is also an opportunity to call attention to the voices and grievances of *other* peoples, the nonhuman animals that frequently populate her poems. That Harjo "had some horses" and returned "home" with the deer and "laughed" with a crow or two—these are not mere poetic tropes, Romantic primitivist fancies. For one thing, in Indigenous cultures—and in Native poetry—there is a characteristic embracing of other animals as close relatives. If N. Scott Momaday claims to be a bear, Harjo is at times a wild deer, now a powerful horse, now a laughing crow. Moreover, Harjo's birds have a far different feel than most comparable imagery from, say, the British and American literary canon, issuing from a new (or "old") poetic world in which the bird itself is more frequently given the right to be an autonomous, integral being and is able to "speak back" to both human poet and audience. Also, if poetry may indeed be *the* form of human discourse most amenable to a veritable rapprochement of human and nonhuman, the genre in which other animals can best be heard "speaking," that possibility is amplified by the positionality of a Native woman poet whose own liminal status allows an even greater degree of empathy for otherness. And it may be in such poetry that we can best recognize the nonhuman animal in its sheer radical alterity, as in Harjo's "blackbirds who are exactly blackbirds" (*Woman* 29).

This chapter examines two of Harjo's favorite avian images, crows and blackbirds, as crucial "other voices" in her corpus. Crows, especially, often speak for the insurrectionary, culture-questioning Native trickster, and Harjo's birds in general are often vocal irruptions of the nonhuman that turn human discourse on its head. They also voice Harjo's eco-message of interspecies relationality and bioegalitarianism. And if they remind us of the ultimate reality of death, at the same time they tell us a new cultural "story," toward a future posthuman reconnection with the nonhuman.

Of Crows and Blackbirds

Beyond the close connection between other animals and Indigenous cultures for millennia, birds have especially struck a chord in contemporary Native poets, in whose work there is an ornithic chorus that speaks long and loud—and is inextricably involved in *story*: "We are story gatherers. That's what humans do. The bird people . . . are up to the same thing" (*Soul* 104). In the wake of her most renowned horse poems (1983) and deer poems (1990), Harjo can state by the mid-1990s that she considers her "Eagle Poem" (1990) and "The Myth of Blackbirds" (1994) to be two of her very "important" poems (*Spiral* 89); and indeed her poetry is permeated by the flights and calls of animals of another feather. Paradoxically, such birds are often both more starkly alien and *other*, as nonmammals, but also more "real," even more "human," as spokesbirds for an alternative—and Native?—reality. For at last, birds are notable as vocalists; they *speak*, even if humans have a great propensity to understand avian language in very anthropocentric ways. Moreover, in contrast to Harjo's horses and deer, and most other wild animals and human pets that humans love as close associates or admire from a safe distance, birds are more obviously utterly nonhuman, *alien*, and so short-circuit our inveterate mammalo-centrism and anthropocentrism to a greater degree.

In combating the tyranny of anthropocentrism, Harjo's favorite avian "other voices" are those of crows and blackbirds. Audrey Goodman precedes me in pointing out this pair's importance, placing them "in dialogue with the jazz voices that sing through" her poetry's "musical lines," rendering her poems "critical Indigenous disruptions . . . and provocations to planetary thinking" (237). I wish to relate such "disruptions" to a sonic semiotics, a meaning making in sound, that falls—or rises—outside the sphere of human poetics, if such a thing is even possible.

While these two birds are not closely related, taxonomically speaking, they are both usually quite vocal and predominantly feathered in a color traditionally denigrated in Western culture. I would suggest that this dark, feathered body may represent another aspect of Harjo's Native (and trickster) insurgency. And maybe the "uglier" the bird, the better. Peter Heymans speaks of the "ecological effect of alienation" as a positive thing. Rather than reading about cute puppies and butterflies, "we need to experience nature as something that feels both intrusively familiar and shockingly uncanny." Environmental consciousness "is not about being nice to birds because they remind us of ourselves." Instead, "it is about becoming aware of the ugly, dysfunctional and inhuman quality of the things we thought we knew" (59–60). Harjo's crows, ravens, and blackbirds do exactly that.

In fact, Harjo's best poetic gestures embrace *this* world, an immanent material realm that has blackbirds in it. The main mythic element in "The Myth of Blackbirds" involves Harjo's ancestors, "the spirits of relatives," who are still present even in the "brutal city" and "spiral of power" that is Washington, DC. In Harjo's worldview, the "white deer intersect with the wisdom of the hunter of grace," and "Horses wheel toward the morning star" in their usual mythopoeic fashion; but the blackbirds remain stubbornly material. The distinctive wings of red-winged blackbirds are described, deliciously, as "the beauty of scarlet licked with yellow" (fig. 10.1). And the poem's concluding thanks go out not only to the poet's "ancestors" of tribal memory but also to the "blackbirds who are exactly blackbirds" (*Woman* 28–29). One is immediately reminded of the various crows in Harjo's poetry that, as we shall see, refuse facile assimilation into her poetic discourse, remaining obstinately existential.

Harjo's blackbirds thus eventually become birds of greater import and suggestiveness, serving as avian companions in ways not unrelated to her trickster crows. In her poem "Transformations," Harjo says, "When I think of early winter I think of a blackbird laughing in the frozen air"; and as much as one wants to simply read "blackbird" as a substitute for (the much larger) "crow" here, the birding reader knows that red-wings are prone to singing their "conk-er-ee" songs late in the season. "This poem could be . . . . a blackbird, laughing," Harjo continues, again giving the blackbird crow-trickster

Figure 10.1. Red-winged blackbird. North Lake Basin Wildlife Management Area, May 29, 2015. Photo by the author.

attributes. The poem's very title, too, is a tribute to other animal species. Positive transformations are possible "if you have the right words, the right meanings, buried in that tender place where the most precious animals live" (*In Mad* 59). What sets Harjo apart is her insistence that the truly "right meanings" are usually beyond the realm of human words.

Corvid Tricksters

Harjo's crows, as alluded to previously, are often culture-critiquing trickster figures, in accord with the poet's tribal background. Kenneth Lincoln characterizes them as "black trickster birds who rave truth" (364). But the "fool crow" in Harjo's "My house is the red earth" is more than some wild raver; "picking through trash in the corral," he

> *understands the center of the world as greasy scraps of fat. Just ask him. He doesn't have to say that the earth has turned scarlet through fierce belief, after centuries of heartbreak and laughter—he perches on the blue bowl of the sky, and laughs. (Secrets 2)*

This crow-centric perspective is one that views life as a tragicomedy, the tragedy of which has notably included the scourge of colonialist conquest spurred by religious fervor. In the poet's "quest" to represent the unpleasant facts of Native history and mundane materiality, her "crows lead the way." Crows allow Harjo "to keep her sense of humor and her perspective," allowing "her imagination" to constantly remain in "shimmering motion, like crows, accepting risk in the face of the necessity to witness and contend with human destruction" (Goodman 245).

The crow's role as Native trickster necessarily entails a bundle of contradictions. The trickster is thus aligned with Mikhail Bakhtin's theorization of the "carnivalesque," with its ambivalent joining of "blessing and curse . . . stupidity and wisdom." It is a sharp-tongued voice of profane "blasphemies," of "obscenities linked with the reproductive power of the earth and the body" (126, 123). Harjo speaks of the Cherokee trickster in a very similar vein: "It's always walking that line between the sacred and the profane—the trickster is always about the duality between here and there, sun and moon, sky and earth," embodying both "the sublime and the ridiculous" (*Soul* 70).

Corvids like crows, ravens, and jays also make great trickster figures because of their intelligence. I would qualify this by noting that even recent science regarding crow intelligence is quite anthropocentric—consisting largely of discoveries that crows' "intelligence" is much like *ours*, with their elaborate language skills, problem-solving abilities, facial recognition, sociality, and so on (fig. 10.2). But their evolutionary success in these regards cannot be denied. Recent research has found that crows "are self-aware and . . . conscious" (Schulze-Makuch). Indeed, via a different evolutionary path, corvid intelligence may be closer to that of primates than most mammalian intelligence is: "The total number of neurons in crows (about 1.5 billion) is about the same as in some monkey species. But because they are more tightly packed, communication between the neurons seems to be better, and the overall intelligence of crows may be closer to that of Great Apes" (Schulze-Makuch). Such intelligence, sociability—and accompanying verbosity—qualify corvids as tricksters par excellence.

J. Scott Bryson reads the crow-trickster as "'a kind of renegade,'" whose persona the poet readily adopts. The crow "offers what Harjo calls the 'laughter of absolute sanity that might sound like someone insane' but" which "is actually 'the voice of sense,'" so Harjo "often looks to the trickster for a

Figure 10.2. Common raven. Valley Wells Rest Area, June 3, 2010. Photo by the author.

model of how to respond" to a contemporary world gone mad (64). This is how Harjo enlists the crow in her poem "Trickster," which aligns the laughing crow with her own "fool" of a Native poet-persona:

> *Crow, in the new snow.*
> *You caw, caw*
> *like crazy.*
> *Laugh.*
> *Because you know I'm a fool*
> *too, like you*
> *skimming over the thin ice*
> *to the war going on*
> *all over the world. (In Mad 13)*

By identifying with another species, Harjo becomes a trickster herself, "disrupting traditional Western ways of thinking and knowing by questioning reality" (Bryson 64). Her poetic persona embraces this trickster arationality because it's in her very name; Harjo, she tells us, is "an Anglicized version of *Hadjo*, meaning 'so brave you're crazy'" (*Map* 61).

But the "war going on" must also be read as an ecological and Native one of animal and Indigenous survival, not the insane wars waged in the name of false political and religious ideologies about other imaginary worlds that are mere covers for blood lust and capitalist domination. Harjo's own politics, in this regard, is *for the birds*. In the geopolitics of the former Cold War, for instance, she was

> *. . . loyal to neither [the US or USSR],*
>
> *only to the birds who fly over, laugh at the ridiculous*
> *ways of humans, know wars destroy dreams, divide the*
>
> *country inside us. (In Mad 60)*

And that war, again, is one that has long been waged against both Indigenous humans and other species. The human Native has been othered as some natured, subhuman, savage, and bestial entity whose preferred status is one of extirpation; likewise, the eagle and crow, and so on, have been othered as sheer *animals*, both idolized and reviled, above all, for their "asocial," extra-human status, and who have thus been deemed better off dead, too, unless redeemed by some (*savage*) "nobility" that has been certified, finally, by the great gods of Western iconography.

Harjo's own corvid song that ends "The Path to the Milky Way Leads through Los Angeles" is a reaction against such othering:

> *I'd rather understand how to sing from a crow*
> *who was never good at singing or much of anything*
> *but finding gold in the trash of humans. (Map 45)*

Goodman notes that the crow "motif" is striking in that "crows rarely rate as muses": "They caw rather than sing." They "are big and not especially graceful, helpful to humans primarily as consumers of carrion and other garbage"; and yet they "reappear throughout Harjo's work, unfazed, unflappable" (44). Harjo's crow is just such a trickster trash collector, a discordant voice that would nevertheless "sing," and even be a better judge of true value than its human confreres. It's as if the Native writer has little choice but to (re)invoke his/her kinship with an abject avian species—and to "sing back" (however dissonantly) as that avian Other, as Harjo does here.

But this passage also recycles, seriously or not, a quite anthropocentric aesthetic of sound, in which the crow's "caw" is sheer cacophony compared to, say, the whistling of a Baltimore oriole; and the corvid's general outsider-with-greater-vision status still issues, in good part, from Harjo's political perception of her own status as a poet of Indigenous otherness and outrage. Harjo exemplifies a counter-discursive response to false Western imaginings about both Indian and bird indigeneities. Her various images of crows and eagles and blackbirds have an intrinsic importance in her corpus, not only as living parts of that natural world that she would champion and "mythically" reinvigorate but also as real individual Others like her, repressed and belittled afterthoughts in the psyche of Western civilization. However unfortunate the historico-political rationale for this connection between Indian and bird, it at least provides an immediate vital link between human and "animal." And it may well be through this othered-as-animal Native that a closer connection between human and animal, words and birds, can best be perceived.

Indeed, women and Natives and other species make up a triumvirate of abjectness and therefore affinity. For Native women writers to protest the othering of this threefold alterity is only—*natural*. One might even argue that the Native woman ecofeminist is *the* renascent Native (and eco-)voice today. As Harjo herself claims, "There is a new language coming about"— "from the women" (*Spiral* 63–64). Harjo's critique of the Garden of Eden myth is telling: "It excludes the power of the snake, the power of women, and the power of the earth mother. And it's a crow who comes back and puzzles over the story and finds a different conclusion. And notice I link 'who' with crow, not 'that' or 'it.' A relationship is made here, not the one of Adam dominating the world" (*Soul* 17). The centrality of the crow in Harjo's eco-aesthetic needs no stronger testimony.

A Native/corvid alliance and resistance must include a deconstruction of the Western division of birds into good-versus-evil categories. Inhabiting the *in*human shadow of darkness, and forebodingly colored in archetypal black, corvids have long been associated with the occult, evil, and death. However, this demonization of the corvid is belied, through an Indigenous lens, by examining the crow's role in Plains Indian oral literature, where the bird is actually closer to the eagle, as spiritual intermediary. In one Lakota Ghost Dance, a crow succeeds an eagle, in fact, performing the same gift of bearing "the message to the tribe, / The father says so, the father says so." In a Cheyenne Ghost Dance lyric, the crow's comparable mythic magnificence becomes manifest: "The crow—*Ehe'eye'!* . . . I saw him when he flew down. . . . He has renewed our life" (qtd. in Mooney 1072, 1035). Indeed, the foremost interpreter of the Ghost Dance songs, James Mooney, deemed the crow to be *the* "sacred bird of the Ghost dance," the veritable "lord of the new spirit world" who would be ushered in by the dance, to revitalize Native cultures (978, 1035).

Maybe the best "trickster" conclusion here is that the crow-trickster is an (anti-)concept that negates all binaries, of good versus evil, human versus nonhuman, and so on. Harjo's crow is more than merely one Native writer's deconstruction of some essential avian evil and otherness—an evil only "universal" to Western civilization; indeed, crows, ravens, and even vultures lack such sinister connotations in many Native cultures. Notably, Harjo's garbage-picking crow is a lowly *scavenger*—allowing yet another Indigenous intervention, reflecting a general positive regard among Indigenous writers for crows (and vultures, etc.) as ecological cleansers of the physical environment; here one might say that they are capable of a human psychological cleansing, too, of a Western collective psyche so inveterately full of human "trash" in the form of false projections and ideologies and binaries. Harjo and her crows are still rummaging through the trash together—and laughing (fig. 10.3).

Other Languages

Having proposed avian voices as a viable part of the world's meaning-making languages, I would claim the crow-trickster's laughter and other such "voices" as non-human-linguistic irruptions of "animality" that would turn human linguistic signification on its head. Harjo's awareness of a "spectrum of other languages" includes such odd tongues as "cloud language, cricket singing talk, and the melodic whir of hummingbirds" (*Spiral* 99). This language spectrum also—most notably—includes the ornery voices of corvids,

Figure 10.3. Fish crows. Roaring River State Park, July 5, 2018. Photo by the author.

a more radical alterity than just having another animal "speak" for a human poet's particular politics or ethnicity. It involves an epistemological skepticism toward—and yet a deep regard for—semiotics that are beyond human words; and the laughter of the trickster surely extends beyond those words. It might even be a useful heuristic to propose corvid language as another unknown and untranslatable "Indigenous" language that must finally be left its mystery, as "the absolute alterity" of the "absolute other" (Derrida 11). Such untranslatability, paradoxically, may well be a positive thing. According to Heymans, the "representation" of the animal other's "unrepresentability might be the closest we come to understanding it" (33–34). Heymans's discussion of an "ecological sublime" is illuminating here, as "that moment when the animal appeared too inhumanly different to be conceptualised" (23). Such a conceptualization might even be a worthy "tool of environmental emancipation," a liberating realization that "it is not reason that transcends nature, but nature that overpowers . . . reason." This negative recognition of the "failure of representation . . . would respect nature's unfathomable otherness" (27–28).

Harjo's crows do partake in human-Native-trickster discourse, as we have already seen. If Harjo's crows "speak to the persistence, patience, acceptance, and humor required to survive in and transform the earth," her persona speaks "through" the bird, reinventing her white "enemy's poetic language" (Goodman 253, 246) and writing back against Western racist colonialism and speciesism. While Harjo's greatest strength is her appreciation of the sheer non-human-verbal nature of "bird talk," she is also thoroughly at home with the avian interlocutors of Native tradition, in an abiding belief that other species can actually communicate with humankind. This is far different—and more eco-therapeutic—than the converse Western imperialist Self's assumption that other species (and other races) are mere dumb animals.

But Harjo's nonhuman voices are most redemptive when they transcend human linguistics as much as possible. As her 2005 interview with *Triplopia* begins, "Joy Harjo knows noise." Her writing is "rich in the auditory imagery of dogs barking, the ground speaking, and the moon playing the horn. And yet, sounds do much more than play to the senses in Harjo's poetry" (*Soul* 3). These sounds include crows cawing, of course, and mere sensory appeal is truly the least of the import of such sounds. For fundamentally, Harjo's poetry is a concerted articulation of the inarticulate, the raw caw of the crow rather than the verbose squawking of a literary scholar like me. Even the crow's purported laughter, however much supported by Indigenous trickster hermeneutics, is still an anthropocentric misreading and *use* of the bird's sound per se. One is more encouraged by statements elsewhere in Harjo's work that more decidedly privilege the superiority of nonlinguistic communication, a faith in the value of sonic semiotics per se, beyond human signification. Indeed, Harjo's is an ironic poetic cosmos in which the truth issues most from that place where "There are no words, only sounds" (*In Mad* 36). Her home ground, the "red earth" of Oklahoma, is a "center of the world" that remains beyond language: "Words cannot construct it, for there are some sounds left to sacred wordless form" (*Secrets* 2). This linguistic truth came to her at an early age, she claims; even as a young girl, the sound of music had already led her to an understanding of "the failings of language, before I could speak" (*Spiral* 102). Now, as a poet writing words, she can still claim the following: "All poets / understand the final uselessness of words" (*In Mad* 21).

Of course, that there are languages of many other species beyond the scientifically ascertainable semantics of chimps and dolphins is another common refrain in eco-scholarship. The faux binary of human "culture versus nature" depends "upon an unexamined premise that the social production of human behavior is entirely distinct from the means by which the 'lower animals' learn to hunt, hide, play, and fight." In truth, "many animals . . . are already fully involved in semiotic exchange" (McKusick 16, 17). Bird songs, in sum, are language, too—in case there were no Joy Harjo to remind us of this.

But Harjo's avian interventions also entail a philosophy of language that transcends both naturalism and the Native. This is most famously embodied in her poem about another bird, one whose literal verbal skills aren't even mentioned. "Eagle Poem" epitomizes her redefinition of language, her belief that human language is but one expression of nature's art, that other species' behaviors are "chants" for the cosmos, too, in the truest posthuman sense. In "Eagle Poem," language is said to transcend even sound. There are "languages / That aren't always sound but other / Circles of motion"—epitomized in the eagle's flight (fig. 10.4), circling "in blue sky," a pattern that "swept our hearts clean / With sacred wings" (*In Mad* 65). This is the psychic cleansing that occurs with a greater awareness of eco-relatedness. The eagle's circling is an Indigenous "sacred hoop," and Harjo's *real*-ization of the circle is also an inner psychological awakening, a new conscious awareness of biological cycles, a knowledge that we live

and die soon within a
True circle of motion,
Like eagle rounding out the morning
inside us. (65)

If the "Eagle Poem" is "a prayer for our continuance" (*Spiral* 123), that "continuance" is both an ecological one, emblematized in the eagle's flight, and a "spiritual" one, an eco-therapeutic internalization of all that the "circle" and "sacred wings" entail. For our current century of imminent eco-apocalypse, there may be no better *language*, this translation of the bird by the Native, for us to hear. For Harjo, then, language can reside in silent life patterns, and in inarticulate sounds. Significantly, Harjo's semiotic eagle is closely related to her crows, for, as she says in another poem, even human existence is embodied "not in words, but in the motion / set off by . . . the simple flight of crow" (*She Had* 54).

As for the "language of the enemy," Harjo provides a radical critique and reinvention. The English language itself, for starters, is "a male language, not" sufficiently "tribal" (*Spiral* 69). The poet's great "frustration with" such language "stems from anger with the colonization process in which the English language was a vicious tool" (99). Moreover, Euro-languages are implicated in—and indeed, create and perpetuate—a worldview that is chronically anthropocentric, that denies other animals any discursive reciprocity. If Harjo imagines "a language that would break through the mortar and concrete of colonizing languages" (Adamson 126–27), it is to enact a reconnection with the natural world, with the real animal. Harjo's characteristic "return" to her Native "mythic world" is an antidote to a "modern alienation" that "she attempts to bypass by going back," that is, "recovering a worldview that is mindful of" the environment (Bryson 46)—in sum, a "return" to an eco-relational worldview, whether one dubs it "Native" or not.

In the title poem of *A Map to the Next World*, Harjo's instructions include a rediscovery of "the language of the land." This includes a reappreciation of the ornithic: "We no longer know the names of the birds here, how to speak to them by their personal names." But our "relatives," the other species, are still here, waiting, in spite of ourselves: "They have never left us; we have abandoned them for science" (19–20)—for Western rationalism, enforced by Western language, within a worldview that prevents us from hearing other languages. Harjo is adamant that, as the title of another poem

Figure 10.4. Bald eagle. Pawnee Lake State Recreation Area, February 8, 2023. Photo by the author.

in *Map* declares, "humans aren't the only makers of poetry": "It's not just humans who sing for rain, make poetry as commentary on the meaning of life. / We aren't the only creatures, or the most likely to succeed" (112). One might well listen to a crow, or a meadowlark, and hear beyond some projected trickster laughter or some pristine paean to spring. The avian auditory that our ears might be opened to is beyond or before the human verbal—it is rhythm, sound in motion, repetition, "poetic animality"—and it requires attending to. And—following Harjo—this entails a redefinition of *language* itself, as in the sheer phatic brilliance of a cardinal's song, or in the ever-changing semiotic patterns of migrating geese in flight; birdsong, especially, is as much "poetry" as the so-called nonsense vocables of Indigenous ceremonial chants.

Growing up, Harjo heard the "influences of 'many musics'"—including country, gospel, Mvskoke stomp dance songs, jazz, and blues (Lincoln 260)—and, I would add, corvid. Harjo's privileging of "no words, only sounds" culminates in her poem "Bird" and her conclusion regarding "the final uselessness of words"—an ironic statement for any poet to utter. The poem is not about a bird, but rather Bird, Charlie Parker, bebop master of those marvelously "convoluted scales" that make Harjo *somewhat* regret her verbal art form: "inside this poem I can't play a horn, hijack a plane to / somewhere where music" can make her "nerve endings dangle" (*In Mad* 21). As a musician herself, Harjo is quite comfortable privileging music over poetry, since the former "doesn't have the added boundary of words" (*Spiral* 101). And her corvid laughter is another fertile effort to escape this boundary. Championing such nonverbal expressions is much more than an acknowledgment of the inability to express the nonhuman in human discourse. To say that "Harjo makes clear that language, even when used to create poems, is ultimately futile" (Bryson 68) is only half the story; the other half includes my argument here that corvids and other animals do "speak" in Harjo's corpus, as much as that is possible.

In the poem "Grace," Harjo plays trickster—like "Coyote, like Rabbit"—as she and a friend survive an Iowa City winter "with laughter," until they finally find "grace," by "once again" understanding "the talk of animals." Fittingly, by poem's end, she is "still crazy," like a crow (*In Mad* 1). To speak more generally, the many speaking animals of contemporary Native poetry are no doubt concerted attempts to recover this state of animal "grace," a gesture reflecting an ongoing faith that such nonhuman "talk" still greatly matters. If the crow is "calling us back," then, it is also calling us toward a more eco-therapeutic future.

We Are All Related

If there is a humanly meaningful "message" in Harjo's corvid laughter, a story discernible in all this animal caw and chatter, it is that of interspecies relationality and bioegalitarianism. Vine Deloria Jr. has repeatedly emphasized the vast difference between the Western and Native worldviews in their approaches to other species, apparent in a characteristic Indigenous way of seeing and being: "In the religious world of most tribes, birds, animals, and plants compose the 'other peoples' of creation," and "various of these peoples participate in human" ceremonies; in contrast, "non-Indians have engaged in senseless killings of wildlife and utter destruction of plant life and *it is unlikely that they would have understood any effort by other forms of life to communicate*" (208, 209; emphasis mine). Here again Bryson is helpful: "instead of an emphasis on individual subjectivity, Harjo's work emphasizes a radical intersubjectivity, with human and nonhuman figures repeatedly fusing into other bodies and psyches." In fact, "an element of the uncanny is woven throughout Harjo's work: a deer may turn into a woman . . . a bird into the sun" (65). This "transformation motif" is "a vehicle by which to convey the 'continuity' between man and other species and to oppose the colonization of man's animal nature. . . . And, in so doing," Harjo "challenge[s] the concept that man is superior to, or has dominion over, 'other' animals" (Haseltine 85).

Once again, this is not to say that Harjo calls for some simplistic "return of the Native." Harjo is well versed in Western ideas of ecology and environmentalism. Her assertion that "We are linked by leaf, fin, and root" (*How We* 176) smacks as much of Darwin as it does of Mvskoke tradition. And the eco-future that she imagines would be a cross-cultural dialogue: "One day there will be a conversation. Everybody will sit down at the table: Cotton Mather and his people, my people, the eagle, the stone, the plants, . . . all of us. And we will be equal. And everyone's voice will have a place" (*Soul* 75–76). Harjo's poetry goes beyond the expected clichés in its earnest call to "understand the world as a world of consciousnesses, including animal consciousnesses" (80). In terms of bioegalitarianism, Harjo is certain that the "human is not above the bear, nor is Adam [in] naming the bear" (*Spiral* 127), another word arrow against Western religious ideology. Witness, especially, her acknowledgment of other animal lifeways, the self-effacement in her ruminations on a heron:

> *I did not see guilt in his posture, nor did I hear him admonish himself for some failure of the deep or near past, rather he absolutely enjoyed his heron-ness. . . . He had no doubt as to his right to be a heron. . . . But what do I know of herons? I do not know their language or their culture. (Map 90)*

Certainly this is a healthier eco-attitude, to see the crow and heron nations as radically different alterities, to grant them their own working languages, in contrast to rendering them spokesbirds for human ethnic politics. Goodman goes as far as to claim that Harjo's crows "remind readers that the natural, more-than-human world demands recognition. Through her crows, she affirms the radical equality of all beings" (245). Harjo emphasizes this bio-egalitarianism in the poem "Anniversary," one of her many re-visions of a tribal creation myth, in which she celebrates a time when there was "no separation" between bird and human:

> *And then a bird or two were added, the crow of course to*
> *joke about humanity, and then another kind so beautiful*
> *we had to hear them first, before our eyes could be imagined.*
>
> *And it was, we were then—and there was no separation. (Map 106)*

Once again, in Harjo's version of human/bird coevolution, what is heard is prioritized. And "of course" the trickster crow gets special mention.

This interrelationality is also often portrayed via a point-of-view shift, in which the hierarchy of human/nonhuman gets turned on its head. Asked about all the "animals" that show up in her poems—"what do they help you say?"—she replies, "Maybe it's the opposite!" (*Soul* 80). The opposite, presumably, is "what do I help *them* say," implying that they are *speaking* through her as a human conduit, as Harjo asks the reader to consider a diversity of species' ways of knowing. In one of her most stunning examples of such a reversal, the human poet becomes simply an image in the mind of a bird:

> *I will appear in the vision of a dove*
> *who perches on the balcony*
> *of the apartment.*
> *In his translation I am the human with a store*
> *of birdseed. He is the sun.*
> *I am a fruitful planet. (How We 192)*

The reversal here is metaphorically Copernican—a revolutionary (pun intended) attitude in which humans are but satellites to other animal species.

Due in good part to humankind's loss of conscious relationality with the rest of the biosphere, our recent impact on the environment hasn't been a laudable one, in Harjo's view. When her son was born, she "wondered what would happen to us, where humans fit in the evolutionary scale. Were we truly necessary to the survival of the biosphere? . . . What do humans add besides stacks of trash and thoughtlessness?" (*Map* 108). So Harjo calls for a more eco-relational way of thinking. In another poem, Harjo invokes a traditional Mvskoke "harmony" with the natural world in which all species'"worlds are utterly interdependent"; she then thinks "seven generations" into the future—but characteristically expands the definition of those generations: "All of our decisions matter, not just to seven generations and more of human descendants, but to the seven or more plant descendants and animal descendants." But of course, "To understand each other is profound beyond human words" (*Conflict* 55)! We are once again left with the tension of a proactive message of eco-interrelatedness presented in human discourse and a natural reality that lies forever beyond human words.

Death, the Central Topic

In Harjo's poetry, the language of crows is most effective when it is less anthropocentric and traditionally trickster-mythic, when it is more suggestive, mysterious, even more abjectly *other*. In a note to "WHO INVENTED DEATH AND CROWS AND IS THERE ANYTHING WE CAN DO TO CALM THE NOISY CLATTER OF DESTRUCTION?," Harjo claims, "*When I hear crows talking, death is a central topic*" (*Woman* 27). But the statement here is more than the usual association of black carrion eaters with dark death—or even corvid tricksters laughing at mortality. If death is the final mystery, crows also remain a mystery whose language is devoid of pat answers. "What a hard year. We're all dying," the poem begins, "even that crow talking loud and kicking up snow." Perhaps the crow "thinks he can head it [death] off with a little noise, a fight," but then Harjo considers the possibility that death is the bird's very talking point. If so, "That's what I like about crows. . . . They aren't afraid to argue about the inarguable." The crows are subsequently invested with a more mystical quality, acting upon humans like Zen riddles: "We fly into the body and we fly out, changed by the sun, by crows who manipulate the borders of reason." Crows are so downright frustrating in their "arguments," in fact, that Harjo speaks of "the talk of crows getting in the way of poetry"—and maybe human discourse in general—and at last, there is something in her that has "little patience for crows." "WHO INVENTED

DEATH" ends appropriately in an epistemological retreat, the poem concluding with a "question" and "laughter," and a final exasperatingly postmodernist question for the reader—"What do you make of it?" (*Woman* 26–27). This is a final acknowledgment of an alien alterity, strange and disruptive in its otherness, as Harjo makes a graceful retreat from any blithe negative association of crows with death. But the association remains: both corvids and death entail a natural reality beyond human rationality.

Death is—for the Western psyche, at least—perhaps *the* reality (and unconscious complex) most repressed by cultural consciousness; and so its intermittent irruptions into discourse often strike us as uncanny, even sublime, like Harjo's corvids at their best. This explains their discordant incomprehensibility, as "arguments about the inarguable." They are figures that take us out of ego consciousness and our vaunted rationalism—and our blithely unconscious faith that we will never die. They remain questions, not answers:

> *Crows mark the border*
> *Between despair*
> *And joy*
> *They are*
> *Poets of noise—*
> *Needed, because the question*
> *Is too large to fit*
> *One city, one church,*
> *Or one country. (How We 182)*

A recent Harjo poem (2019) begins "I Wonder What You Are Thinking" and develops the thoughts of a "feathered wife" and "feathered husband" in pretty typical anthropocentric fashion. But then, to his wife's queries about what he is thinking, the bird-husband "says nothing— / As he wonders about the careless debris that humans make." To her last exasperated query, "*I wonder what you are thinking*," he replies "Nothing. / I was thinking about the nothing of nothing at all" (*American* 72–73). This is the reply of the mundane and the real, of Wallace Stevens's "Thirteen Ways of Looking at a Blackbird," and of Joy Harjo's crows (fig. 10.5).

So, if some of Harjo's most memorable corvid imagery entails an association of crows and death, it does so in a disturbing, uncanny fashion. "In the

Figure 10.5. Fish crows. Roaring River State Park, July 5, 2018. Photo by the author.

Beautiful Perfume and Stink of the World" is one of Harjo's several accounts of an out-of-body experience, leading at last to a vision of her apocalyptic "fifth world"—Harjo's version/vision of Indigenous and ecological restoration. But her manner of flight is rather startling: *"I had been traveling in the dark, through many worlds, / the four corners of my mat carried by guardians in the shape of crows"* (*Map* 133). The four crows here echo the four directions, four worlds, four winds, and so on, of many Native tribes. But in this poem, at least, the future "fifth world" is not just the utopian fruit of a political-eco-revolution, but a new way of seeing, a cleansing of perception, a new awareness of the material here and now that includes the dirty, the "dark"—and the "cawing, flapping" crow:

And it is all here. Everything that ever was.

The cawing, flapping song of the beautiful dark

In the dark. In the beautiful perfume and stink of the world. (135)

This is the song of sheer materiality, of dust-to-dust death, and of crows, of a "corvid consciousness."

Heymans's ruminations on the "ecological sublime" are relevant here: although "songbirds are typically glorified . . . they are actually rather annoying creatures. Their songs render poets conscious of their mortality" and also "underline their alienation from" the natural world (29–30). This "annoyance" is even greater when "trash" birds like blackbirds and crows flock into a poem. Harjo's dark birds remind us of our mortality and our disconnect from materiality, and of the failure of our language and consciousness to even fathom such a reality. But this is ultimately a good thing, for such a "sublime subversion of our tenacious belief in human exceptionality can have an intensely liberating outcome. Once we no longer cling to our special status in nature . . . even the prospect of our own death becomes bearable." Such an awakening "can release us from the narcissism that makes us such tragically frustrated figures, fretting alone at the centre of the world" (Heymans 39–40). Harjo's crows break us out of a Western individualism driven by ego pride and invested in personal tragedy. "Corvid consciousness" moves us into a cross-species material realm in which the question "WHO INVENTED DEATH?" becomes a moot point.

Final Flight

I have read Harjo's "fifth world" as the possibility of a radical change in the collective human psyche, a major cultural attitude adjustment regarding other biota. Harjo's "crow talk" is very much about such an expansion of human consciousness, toward a posthumanist reconnection with the nonhuman. But to speak of a return to some aboriginal appreciation of other species in any but a metaphorical fashion would be disingenuous. The new reimagining of the bird that I see currently taking place is really a cross-cultural cross-pollination that includes the contributions of Euro-American poets like Wendell Berry, Gary Snyder, James Wright, W. S. Merwin, and Mary Oliver—in a "revival meeting," if you will, in Harjo's "fifth world." Despite positive signs from, say, the most recent generation of my students, of many signs of an ostensibly burgeoning global "green" consciousness (which are usually immediately, perhaps inevitably, co-opted by Big Capitalism), there is still much work to be done. The trickster-crows in the poetry of Harjo are still laughing at all our human-centric hubris, perhaps (and hopefully) to signal its end. But if the "fifth world" doesn't arrive, they will then be chortling at humankind in toto, with a more tragic—indeed, sinister and death-rattling—laughter.

In a 2004 essay, Harjo comments on the deep intersection of her art and political-environmental reality:

> I used to imagine writing as a ladder . . . from the blind world into the knowing world, but now to imagine a ladder means to imagine a land or a house on which to secure a ladder. For many of us in these lands now called America, imagining this place has been a tricky feat, because there is no place that hasn't or won't get stolen, polluted, or destroyed, and . . . the foundation is shaky, because . . . the country was founded on violent theft. . . . Maybe the ultimate purpose of literature is to humble us to our knees, to that know-nothing place. Maybe we . . . are a story gone awry, with the Great Storyteller frantically trying out different endings. . . . We need new songs, new stories. (*Soul* 134)

Harjo's crows and blackbirds are a germinal aspect of her own "new songs" and "stories." And "that know-nothing place" seems especially appropriate for an author whose birds continually make us question what humankind thinks we can say or know. "Caw. Caw."

Works Cited

Adamson, Joni. "And the Ground Spoke: Joy Harjo and the Struggle for a Land-Based Language." In *American Indian Literature, Environmental Justice, and Ecocriticism*, by Joni Adamson, 116–27. Tucson: University of Arizona Press, 2001.

Bakhtin, Mikhail. *Problems of Dostoevsky's Poetics*. Translated by Caryl Emerson. Minneapolis: University of Minnesota Press, 1984.

Bryson, J. Scott. *The West Side of Any Mountain: Place, Space, and Ecopoetry*. Iowa City: University of Iowa Press, 2005.

Deloria, Vine, Jr. *For This Land: Writings on Religion in America*. Edited by James Treat. London: Routledge, 1999.

Derrida, Jacques. *The Animal That Therefore I Am*. Edited by Marie-Louise Mallet. Translated by David Wills. New York: Fordham University Press, 2008.

Goodman, Audrey. "Apertures into the Next World." In *A Planetary Lens: The Photo-Poetics of Western Women's Writing*, 233–59. Lincoln: University of Nebraska Press, 2021.

Harjo, Joy. *An American Sunrise: Poems*. New York: Norton, 2019.

———. *Conflict Resolution for Holy Beings: Poems*. New York: Norton, 2015.

———. *How We Became Human: New and Selected Poems: 1975–2001*. New York: Norton, 2002.

———. *In Mad Love and War*. Middletown, CT: Wesleyan University Press, 1990.

———. *A Map to the Next World: Poetry and Tales*. New York: Norton, 2000.

———. *Secrets from the Center of the World*. Photographs by Stephen Strom. Tucson: University of Arizona Press, 1989.

———. *She Had Some Horses*. 3rd ed. New York: Thunder's Mouth Press, 2006.

———. *Soul Talk, Song Language: Conversations with Joy Harjo*. Edited by Tanaya Winder. Middletown, CT: Wesleyan University Press, 2011.

———. *The Spiral of Memory: Interviews*. Edited by Laura Coltelli. Ann Arbor: University of Michigan Press, 1996.

———. *The Woman Who Fell from the Sky: Poems*. New York: Norton, 1994.

Haseltine, Patricia. "Becoming Bear: Transposing the Animal Other in N. Scott Momaday and Joy Harjo." *Concentric: Literary and Cultural Studies* 32, no. 1 (January 2006): 81–106.

Heymans, Peter. *Animality in British Romanticism: The Aesthetics of Species*. New York: Routledge, 2012.

Lincoln, Kenneth. *Sing with the Heart of a Bear: Fusions of Native and American Poetry, 1890–1999*. Berkeley: University of California Press, 2000.

McKusick, James C. *Green Writing: Romanticism and Ecology.* New York: St. Martin's Press, 2000.

Mooney, James. *The Ghost-Dance Religion and the Sioux Outbreak of 1890.* Lincoln: University of Nebraska Press, 1991.

Schulze-Makuch, Dirk. "Crows Are Even Smarter Than We Thought." *Smithsonian Magazine*, February 10, 2021. https://www.smithsonianmag.com/air-space-magazine/crows-are-even-smarter-we-thought-180976970/.

CONTRIBUTORS

Mary McAleer Balkun is professor of English at Seton Hall University. She is the author of *The American Counterfeit: Authenticity and Identity in American Literature and Culture* (2006), coeditor of *The Companion to American Poetry* (2022), coeditor of *Transformative Digital Humanities: Projects, Case Studies, and Challenges* (2020), coeditor of *Women of the Early Americas and the Formation of Empire* (2016), and associate editor of *The Greenwood Encyclopedia of American Poets and Poetry* (2005). She has published essays on early American topics, educational technology, and curricular change. Her current project is *Indelible Connections: Familial Correspondence in the Early United States*.

Thomas C. Gannon is professor of English and of Indigenous Studies at the University of Nebraska–Lincoln. His publications include *Skylark Meets Meadowlark: Reimagining the Bird in British Romantic and Contemporary Native American Literature* (2009) and various articles on the intersection of birds and human discourse (which he has dubbed "ornithicriticism"). His most recent book, *Birding While Indian: A Mixed-Blood Memoir* (2023), is part birding memoir, part cultural critique of the ongoing Christo-Custer colonialism of the Great Plains. He is an enrolled member of the Cheyenne River Sioux Tribe.

Calista McRae is an associate professor at the New Jersey Institute of Technology. She is author of *Lyric as Comedy: The Poetics of Abjection in Postwar America* (2020) and coeditor of *The Selected Letters of John Berryman* (2020).

Aaron M. Moe, an independent scholar, is the author of *Ecocriticism and the Poiesis of Form: Holding on to Proteus* (2019) and *Zoopoetics: Animals and the Making of Poetry* (2014)—along with several chapters and articles on ecopoetics/zoopoetics. In 2021, he published a book of poems, *exhalations*. His creative work can also be found in a leaflet of aphorisms, *Protean Poetics* (2015), in edited collections by Twenty Bellows, and in the anthology *Counterclaims: Poets and Poetries, Talking Back* (2020). He and his family live near the foothills of the Colorado Rockies where they hike, climb, run trails, and write.

Philip Edward Phillips is professor of English Emerita and associate dean of the University Honors College at Middle Tennessee State University. His research interests include nineteenth- and twentieth-century American literature. His chapter on Poe's Ourang-Outang in "The Murders in the Rue Morgue" appears in *Animals in the American Classics* (2022), edited by John Cullen Gruesser. His own edited collection of essays, *Poe and Place*, received the Poe Studies Association's J. Lasley Dameron Award for 2018. He is president of the Poe Studies Association.

Susan L. Roberson is Regents Professor of English at Texas A&M University–Kingsville. Her works include *Antebellum American Women Writers and the Road: American Mobilities* (2011) and *Emerson in His Sermons: A Man-Made Self* (1995). A new collection of essays she edited, *Women across Time, Mujeres a Través del Tiempo: Sixteen Influential Women of South Texas*, was published by TAMU Press in 2022, and another, *Geographies of Travel: Images of America in the Long Nineteenth Century*, was published in 2024. Her scholarly interests include nineteenth-century American literature and American travel writing.

Daniel Cross Turner's books include a poetry collection, *Riding Light* (2024); a monograph, *Southern Crossings: Poetry, Memory, and the Transcultural South* (2013); an anthology featuring seventy-seven poets, *Hard Lines: Rough South Poetry* (2016); an essay collection on the gothic, *Undead Souths: The Gothic and Beyond in Southern Literature and Culture* (2015); and an anthology with fifty poets exploring coastal ecologies, *Coast Lines* (2025). His essays appear in academic journals, including *Mosaic* and *Genre*, and in edited collections by Oxford and Cambridge, and his creative writing appears in *Birmingham Poetry Review*, *Five Points*, and *Literary Matters*. For more, see www.danielcrossturner.com.

Margarida Vale de Gato is associate professor and coordinator of the American Studies program in the School of Arts and Humanities and a researcher in the Centre for English Studies at Universidade de Lisboa. She is the editor, with Emron Esplin, of the volumes *Translated Poe* (2014) and *Anthologizing Poe: Editions, Translations, and (Trans)National Canons* (2020), winner of the J. Lasley Dameron Award. She is the author of the chapter "Poe and Modern(ist) Poetry" for *The Oxford Handbook of Edgar Allan Poe* (2018). A poet and literary translator, she has edited and rendered Poe's complete poetical works into Portuguese and is the author of three poetry collections: *Mulher ao Mar* (2010–23), *Lançamento* (2016), and *Atirar para o Torto* (2021).

Heather Cass White is professor of English at the University of Alabama. She is the editor of the *New Collected Poems of Marianne Moore* (2017). She is the author most recently of *Books Promiscuously Read: Reading as a Way of Life* (2021) and is currently editing the complete correspondence of Elizabeth Bishop and Marianne Moore.

Brian Yothers is professor and chair of English at Saint Louis University and the editor of *Leviathan: A Journal of Melville Studies*. He is the author of *Sacred Uncertainty: Religious Difference and the Shape of Melville's Career* (2015) and *Melville's Mirrors: Literary Criticism and America's Most Elusive Author* (2011), among other books. He is coeditor with Jonathan A. Cook of *Visionary of the Word: Melville and Religion* (2017) and with Harold K. Bush of *Above the American Renaissance* (2018), and editor of Broadview editions of *The Piazza Tales* (2018) and *Benito Cereno* (2019) and of *Billy Budd: Critical Insights* (2017).

INDEX

Ackerman, Jennifer, 87
aestheticization, 167–168, 170, 172
alligator, 55–56
Amherst, Massachusetts, 67, 70, 76, 77, 78
Amherst College, 83, 84
amphibians and reptiles. *See individual species*
angleworm. *See* worm
"animal magnetism," 30
animal studies, 24, 30, 46
animal speech, 23–24, 26–29, 33, 34, 35, 122. *See also* talking birds
animals. *See individual animal types*
animality, 46–66, 181, 184, 191, 209, 214
anthropocentrism, 64, 119, 180, 183, 187, 192, 202, 205, 208, 212, 213, 218, 219
anthropomorphism, 2, 7, 14, 16, 29, 42n5, 158, 192, 193
architecture, 63, 92, 93
albatross, 26
armadillo, 191, 192; *See also* "The Armadillo" (Bishop)
Auden, W. H., 157–58, 164, 174n1
Audubon, John James, 8, 33, 86, 144

Bakhtin, Mikhail, 205
basilisk (plumed), 139
bat, 25, 36, 37, 140
Baym, Nina, 85
bear, 6, 190, 201, 216
behemoth, 4
Berkeley, James, 28
binaries, 209, 212
"biopower," 31
bioegalitarianism, 202, 215, 216, 217
bioregionalism, 85–88

biosemiotics, 47, 48, 52–58
bird songs and calls, 87, 212
birds. *See individual bird types*
Bishop, Elizabeth, 146–147, 157–178; "The Armadillo," 167, 169, 170, 172; "At the Fishhouses," 157, 161; "Cape Breton," 167; "The Fish," 169–72, 175n12, 176n14; "Five Flights Up," 164; "The Last Animal," 160; "Pink Dog," 169; "Roosters," 172–74; "Under the Window: Ouro Prêto," 165–66, 174
blackbird, 7, 70, 202–04, 208, 219, 221, 222
Blake, William, 193
Bliss, Rachel, 191–93
bluebird (eastern), 85
Bly, Robert, 133
bobolink, 70, 78–79, 80
Boggs, Colleen, 3–4, 24, 27, 29, 42n6
Braidotti, Rose, 30
Bradstreet, Anne, 1–22; "A Letter to Her Husband, Absent Upon Public Employment," 14; "Another (I)," 14–15; "Another (II)," 13, 14–16, 18, 19, 20n14; "The Author to Her Book," 3; "Before the Birth of One of Her Children." 12–13; "Contemplations," 9–12, 13, 20n13; "In Reference to Her Children, 23 June 1659," 3, 13, 16, 19; "The Four Elements," 4–8; "The Four Seasons," 7–8; "To My Dear Children," 3; "Verses Upon the Burning of Our House, July 10, 1666," 13
Bryant, William Cullen, 79
Bryson, J. Scott, 205–07, 216
Buell, Lawrence, 68

bull. *See* cow
bugs. *See* insects and arachnids

camel, 6, 107–109
Cameron, Sharon, 76
Carey, Brycchan, 42n7
cat, 95, 107, 137, 138, 139, 160, 193
cattle. *See* cow
Chevalier, Jean, 42n8
chicken, 7, 127–29, 132, 160
Christian symbology, 78–85
Christianity, 85, 107, 108, 173
cock-fighting, 173
Coleridge, Samuel Taylor, 26, 32; *The Rime of the Ancient Mariner*, 26
colonialism, 205, 211
"com-post," 25, 29–36, 36–40
condor, 27, 33, 38
Cooper, Susan Fenimore, 72
coral, 92–94, 97
cow, 6, 8, 49, 119, 140, 160
crane, 6
Crawford, Allen, 53–55
creativity, 70, 75, 77, 88
crow, 46, 70, 201, 202–09, 212, 213, 215, 216, 217, 218, 219–21, 222
"critter," 23–45

Dante, 120, 125, 132n5
"dark ecology," 32, 34, 36
Darwin, Charles, 216
Dayan, Colin, 93
Dayan, Joan, 30
deer, 14, 15, 16, 151, 201, 202, 203, 216
Deloria, Vine, Jr., 215
Derrida, Jacques, 28, 42n2, 46
Dickinson, Emily, 65, 67–91, 95, 107, 119, 132n3; eye malady, 76–77; "Bees are Black, with Gilt Surcingles—" (#1405), 71–72; "Before I got my eye put out" (#327), 67, 76–77; "A Bird came down the Walk—" (#328), 76–77; "The Birds reported from the South—" (#743), 84; "Flowers—Well—if anybody" (#137), 74; "The Gentian weaves her fringes—" (#18), 70–71; "How soft a Caterpillar steps—" (#1448), 72–73; "I heard a Fly buzz—when I died—" (#465), 74, 81–83; "I taste a liquor never brewed" (#214), 107; "The Judge is like the Owl—" (#699), 69–70; "Like Brooms of Steel" (#1252), 74; "The Missing All—prevented me" (#985), 67, 68, 69, 81, 85, 87, 88; "A narrow Fellow in the Grass—" (#986), 71, 74, 81; "New feet within my garden go—" (#99), 84; "Of Being is a Bird" (#653), 74, 76; "A Pang is more conspicuous in Spring" (#1530), 85; "A prompt—executive Bird is the Jay—" (#1177), 72; "The Robin is a Grabriel" (#1483), 75; "The Robin's My Criterion for Tune—" (#285), 85–87; "A Route of Evanescence" (#1463), 73–74; "The Skies can't keep their secret!" (#191), 79–81; "Some keep the Sabbath going to Church—" (#324), 78–79; "Some Rainbow—coming from the Fair!" (#64), 83; "The Spider holds a Silver Ball" (#605), 72; "The Sunrise runs for Both" (#710), 74–75; "Tell all the truth but tell it slant" (#1129), 119, 132n3; "There came a Wind like a Bugle" (#1593), 75; "There is no Frigate like a Book" (#1286), 75; "There's a certain Slant of light" (#258), 76; "Within my Garden, rides a Bird" (#500), 75; poetic method, 87
Dillard, Annie, 75
Dobbin. *See* horse
dog, 68, 75, 138, 160, 162–165, 166, 169, 175n9
domestic standpoint, 74
domestication, 153, 162, 163
dolphin, 212
Dulac, Edmund, 36–37

eagle, 95, 108, 166, 202, 207, 212–213, 214
ecology, 72, 112, 181, 183, 184, 186, 195, 198, 216
Eden, 2, 80, 84, 105, 110–111, 185, 208
elephant, 6, 46, 139, 151–52
Ellerman, Winnifred "Bryher," 150
Eliot, T.S., 136, 137, 150, 154n3
Emerson, Ralph Waldo, 26, 34, 50–51, 59, 79
environment, 1, 2, 68, 75, 77, 102, 142, 146, 192, 195, 213, 217
environmentalism, 216
evolutionary, 180, 190, 192, 205, 218
exploitation, 4, 112, 162, 172, 174, 175n6

Faggen, Robert, 119, 133n12
Faulkner, William, 116, 180
Felstiner, John, 115, 118
Fetterley, Judith, 69
fish and crustaceans: catfish, 182; crawfish, 183; grouper, 169, 171, 175n12; mullet, 14–15, 16; parrot fish, 175n12; pilot fish, 94–95, 100–103, 105; shark, 100–103
fishing, 169–70
Fini, Léonor, 38–41
frigate bird/frigate pelican, 136, 144–146, 147, 151
frog, 83, 119, 123, 124, 132, 133n9, 182
Frost, Elinor, 115, 130
Frost, Lesley, 116, 118
Frost, Robert, 115–135, 136; Derry, New Hampshire, 115, 116, 123, 130, 134n21; England, 117; New England, 115, 116, 117; Pulitzer Prizes, 131, 132n4; *A Boy's Will*, 115, 116; "A Blue Ribbon at Amesbury," 126, 127–128; *Collected Poems*, 117, 132n2, 132n4; "Design," 128–129; "The Exposed Nest," 120, 121–122; *A Further Range* 127, 128, 132n4; "Hyla Brook," 122–124, 126; *In the Clearing*, 155; *Mountain Interval* 116, 121, 122; "The Need of Being Versed in Country Things," 126–127; *New Hampshire*, 120, 126, 132n4; *North of Boston*, 116, 117, 120; "The Oven Bird," 122, 124–126, 133n7; "The Pasture," 117–119, 130–131; "Stopping by Woods on a Snowy Evening," 120; *A Witness Tree*, 132n4; "The Wood-Pile," 120–21

gannet, 95, 112
Garrard, Greg, 43n5
geographic imagination, 68–91
Gheerbrandt, Alain, 42n8
Gohdes, Clarence, 81
Ghost Dance, 209
Gieseking, Jen Jack, 68
glacier, 99
Goodman, Audrey, 202–203, 207–208, 217
goose, 214
Grahame, Kenneth, 137
Graves, John, 83–85
Greenfield, Sayre, 42n7
griffin, 32
Grunes, Marissa, 102–104

Haraway, Donna, 24–26, 31–32, 40, 42n4, 46
Harjo, Joy, 201–224; "Anniversary," 217; "Bird," 215; "Eagle Poem," 202, 212–213; "Grace," 215; "I Wonder What You Are Thinking," 219; "In the Beautiful Perfume and Stink of the World," 219–221; *A Map of the New World*, 213; "My house is the red earth," 204–205; "The Myth of Blackbirds," 202, 203–204; "The Path to the Milky Way Leads Through Los Angeles," 207–208; "Transformations," 203–204; "Trickster," 206–207; "WHO INVENTED DEATH," 218–219
Hardy, Thomas, "A Darkling Thrush," 124–125

hawk, 18, 76, 100
Hemingway, Ernest, 176n14; "Big Two-Hearted River," 170
Heymans, Peter, 203, 211, 221
Higginson, Thomas Wentworth, 81, 87–88
Hitchcock, Edward, 83, 84
Hitchcock, Orra White, 83, 84
Hodgman, Joanna Bailey, 70
Holmes, John, 124–125
Homo sapiens, 130
horse, 6, 109–110, 119–120, 132n5, 166, 201–202
human-animal interdependence, 182–183
human beings: and animal traits, 2, 23; exceptionalism, 31; inconsistency, 166; language, 195, 212; reason, 23–24, 192; speech, 23–24, 26–29, 192
hummingbird, 70, 73–74, 75, 209
"humusities," 32
hybrid, 24, 34, 36, 39–40, 181, 188, 190, 191–196
hyena, 196
hyla. *See* frog
hyperobjects, 47–48, 50, 65n2

iceberg, 102–04
insects: bee, 7, 56, 68, 71–72, 74, 75, 83–84; butterfly, 34, 38, 70–71, 83, 104–106, 110, 139; caterpillar, 72–73, 143; cricket, 9–10, 12, 62, 209; dragonfly, 12, 47, 62–64; fly/maggot, 68, 71, 74, 81–83, 181, 188, 189; grasshopper, 9–10, 12, 62; katydid, 62; locust, 62, 142, 146; moth, 34, 128–130
intentionality, 23, 42n2, 93
Islam, 110

Jarrell, Randall, 117, 134n20, 143–144
jay, 70, 72, 73, 83, 205
jerboa, 137, 141–144, 145, 146, 147, 148, 151, 152
Johnson, Barbara, 35
Jonik, Michael, 93
Judaism, 110

Keats, John, "Ode to a Nightingale," 132n7
Kendall, Tim, 121, 127, 129–30, 134n18
kitten. *See* cat
kiwi, 139
Klu Klux Klan, 195
Komanyakaa, Yusef, 179–200; "Another Kind of Night," 195–196; *Dien Cai Dau* 181, 183; "The Edge," 186–187; "The Four Evangelists," 181; "Greenness Taller Than Gods," 184–186; "The Leopard," 193–194; "The Millpond," 181–183; *Night Animals*, 181, 191–196; "Nightriders," 195–196; "Ode to the Maggot," 187–189; "Slime Molds," 181, 188–190, 198n3; *Talking Dirty to the Gods*, 181, 187–190
Kress, Gunther, 23
Kupka, František, 37–38

lark, 6, 214
leopard, 4, 69, 193–194, 197
Lincoln, Kenneth, 204
lion, 6, 13, 140, 190
Lispector, Clarice, "A Hen," 175n4
Lowell, Amy, 117, 131
Lowell, Robert, 167, 175n12
Lyman, Joseph, 76
Lynch, Tom, 69

Mabbott, Thomas Ollive, 28, 34
Madison, Robert D., 96, 99
Magistrale, Tony, 37
mammals. *See individual species*
masculinity, 173, 174
McClatchy, J.D., 78
"meditation on the creatures," 9, 13
Melville, Herman, 65n4, 92–114; "The Berg," 94, 99, 102–104; "Butterfly Ditty," 95, 104–106; *Clarel, A Poem and Pilgrimage in the Holy Land*, 95, 101, 107–110; "The Haglets," 94, 96–100, 102, 104; *John Marr*

INDEX | 233

and Other Sailors, 94, 96, 100, 102, 111, 112; "The Maldive Shark," 94, 100–102; "The Man o' War Hawk," 94, 100; *Moby-Dick*, 65n4, 94, 95, 96, 98, 191, 108, 110–111, 112; "Pebbles," 95, 111–112; "Shiloh," 108; "Venice," 92–94; *Weeds and Wildings*, 95, 104
metamorphosis, 33, 34, 38, 73, 180, 190, 192, 197n2
Middle Passage, 195
Milne, Anne, 42n7
Milton, John, 128, 133n16, 133n17
mobility, 68, 69, 75, 88
mockingbird, 191, 192
monkey, 184–185, 198n5, 205
monstrosity, 35
Mooney, James, 209
Moore, Marianne, 136–156, 158, 175n6; "Black Earth," 151–152; "The Buffalo," 136; *Collected Poems*, 136, 150; "Diligence Is to Magic as Progress Is to Flight," 152–153; "Elephants," 151–153; "The Frigate Pelican," 136, 144–146, 147; "The Jerboa," 136, 141–144; "The Pangolin," 147–150, 154n8; *The Pangolin and Other Verse*, 147; "Peter," 139; "Pigeons," 154n3; "The Plumet Basilisk," 136; "Poetry," 139, 140 ; *Selected Poems*, 137, 150; "The Student," 140; *What Are Years*, 150
"more-than-human," 29, 35, 37, 217
Morton, Timothy, 32, 34–35, 48, 50, 198n4
Mother Carey's Chicken, 95, 108
murmuration, 56

Native Americans/Indigenous people, 2, 176n16, 201–224
natura natarans, 192–193
nature: Darwinian view of, 83, 128, 130; transcendentalist view of, 26, 79
nature writing, 68
nonhuman consciousness, 24, 180, 216

nonhuman language, 27, 35. *See also* talking birds

O'Brien, Timothy, 130
obliviousness, 159, 167
Oerlemans, Onno, 120–121
Olney, James, 69
orangutan, 23, 27, 42n6, 43n10
ostrich, 6, 139, 141
othering, 34, 207–208
ovenbird, 124, 125
owl, 69, 195

Palestine, 107–108
pangolin, 147–151, 154n8
panther, 4
Parini, Jay, 116–117, 128, 133n15
paroquet, 27
parrot, 27
partridge, 6
peacock, 6, 38
penguin, 140
Perkins, William, *Treatise of Man's Imagination*, 9
pig, 159
pigeon, 108–10
Phillips, Philip Edward, 42n6
pintado, 95, 112
Philomel(a), 11, 122n6
phoebe, 70, 127
phoenix, 6
platypus, 191–192
Poe, Edgar Allan, 23–45, 132n7; "Al Aaraaf," 25; *Arthur Gordon Pym*, 23; "The Bells," 35; "The Black Cat," 29; "The City in the Sea," 36; "The Conqueror Worm," 29, 33–34, 36, 37; "Fairy-Land," 26; "The Gold-Bug," 33; "The Haunted Palace," 36; "Instinct vs. Reason: A Black Cat," 24; "Ligeia," 29, 34, 38; "Maelzel's Chess Player," 28; "The Murders in the Rue Morgue," 23, 43n10; "Notes upon English Verse," 34; "Oh, Tempora! Oh, Mores!"

24; "The Philosophy of Composition," 23, 38; "The Raven," 23–25, 28–33, 35, 39–40, 132n7; "Romance," 27; "Sonnet—To Science," 33; *Tales of the Grotesque and the Arabesque*, 33; "To—[Elmira Royster]," 28; "Ulalume," 33
poetic form, 148
Poirier, Richard, 116, 120, 132n2
post-Darwinian, 119, 126–130
posthuman, 28, 30, 34, 202, 212
posthumanist, 28, 30, 31, 221
Pound, Ezra, 136, 150
projection, 172–174, 176n16
Puritans, 2, 3, 9
Pryse, Majorie, 69

Rasula, Jed, 32, 43n9
raven, 24, 27, 28, 29, 30, 31, 33, 39, 40, 42, 48, 203, 205, 209
regional writing, 69, 71
religion, 69, 79, 81, 83, 88, 126, 133n12. *See also* Christianity
robin, 70, 72, 83–84, 85–87

Schuman, Jo Miles, 70
seal, 94, 104, 157–158, 159
Sewell, Franklane, 127
Sewall, Richard, 76, 83
shearwater, 95–96, 99
Shelley, Percy Bysshe, "To a Skylark," 132n7
Shurr, William, 94, 95–96
Skinner, Jonathan, 70
skunk, 139
singing, 10, 28, 87–88, 124, 157, 182, 186, 203, 207, 209
Slayton, Jessica, 37
slime molds, 181, 188–190
Smith, Virginia, 115, 132n1
snake, 62, 68, 71, 74, 81, 82, 83, 184–186, 196, 208
spider, 46, 47, 48, 52, 57–60, 61, 62, 64, 71, 128–130, 184–185
starling, 56

Stein, William Bysshe, 94, 100
Stevens, Wallace, 136, 154n6, 219
Stevenson, Anne, 158, 176n14
stork, 6
Stowe, Harriet Beecher, *Uncle Tom's Cabin*, 83
swallow, 108, 144, 165, 174
Sweeney, Susan Elizabeth, 33–34, 38

talking birds, 27–28
Theocritus, 117
tiger, 6, 140, 194
Thoreau, Henry David, 68, 105–106, 115; *Walden*, 65n4, 115
thrush, 6–8, 124
toucan, 160, 162–163
transcendentalism, 26, 79
trickster, 80, 202–203, 204–209, 211, 212, 214–215, 217, 218, 220
Trilling, Lionel, 128, 130, 134n20
Tuan, Yi-Fu, 69
turtledove, 14–15, 17, 35

unicorn, 4

van Leewen, Theo, 23
vegetarianism, 160
Vergil, 117
Vietnam, 179, 181, 183–187, 197n1
vulture, 13, 25, 33, 99, 186, 209

Walker, Catherine, 102–104
waste, hazardous, 161, 183
Wetmore, Alex, 27–28
whale, 46, 62, 94–95, 111, 161–162
whaling, 161–162, 175n5
Wheeler, Wendy, 52–53
Whitman, Walt, 32, 46–66, 189–190; clef of the universes, 48–50, 52, 57, 62, 63, 64; journeywork, 49–50, 52, 54, 57, 61–62, 63; original energy, 49, 50–52, 54–55, 56–57, 59–60, 61, 64; sex, 47, 57, 61; *Leaves of Grass*, 47, 48, 49, 50–51, 55, 56, 57, 64, 65n1, 65n4,

189; "A Backward Glance o'er Travel'd Roads," 47, 48; "The Dalliance of the Eagles," 48, 57, 60–62, 64; "A Noiseless, Patient Spider," 48, 57–60; "At the Beach at Night Alone," 48; "Clef Poem," 49; *Song of Myself*, 51–52, 53–55, 57. *See also* animality
white Wyandotte. *See* chicken
Williams, William Carlos, 136
Woodbridge, John, "Epistle to the Reader," 2–3
wolf, 6, 190

Wolfe, Cary, 30
Wolff, Cynthia Griffin, 76, 81, 84
worm, 24, 29, 33, 34, 35, 36, 37, 38, 40, 77, 79, 92–93
wren, 6, 49, 70

zebu. *See* cow
Ziser, Michael, 24, 29, 33, 35
zombie, 36
zoomorphism, 2, 14, 16, 19n2, 42n5
"zoosemiotics," 24, 29